MODERN INTERVENTION TOOLS FOR REHABILITATION

MODERN INTERVENTION TOOLS FOR REHABILITATION

MEENA GUPTA

Amity Institute of Physiotherapy, Amity University, Uttar Pradesh, India

DINESH BHATIA

Department of Biomedical Engineering, North Eastern Hill University, Shillong, Meghalaya, India

PRAKASH KUMAR

Amity Institute of Occupational Therapy, Amity University, Uttar Pradesh, India

ACADEMIC PRESS

An imprint of Elsevier

elsevier.com/books-and-journals

Academic Press is an imprint of Elsevier

125 London Wall, London EC2Y 5AS, United Kingdom
525 B Street, Suite 1650, San Diego, CA 92101, United States
50 Hampshire Street, 5th Floor, Cambridge, MA 02139, United States
The Boulevard, Langford Lane, Kidlington, Oxford OX5 1GB, United Kingdom

Notices
Knowledge and best practice in this field are constantly changing. As new research and
experience broaden our understanding, changes in research methods, professional
practices, or medical treatment may become necessary.

Practitioners and researchers must always rely on their own experience and knowledge in
evaluating and using any information, methods, compounds, or experiments described
herein. In using such information or methods they should be mindful of their own safety
and the safety of others, including parties for whom they have a professional responsibility.

To the fullest extent of the law, neither the Publisher nor the authors, contributors, or
editors, assume any liability for any injury and/or damage to persons or property as a
matter of products liability, negligence or otherwise, or from any use or operation of any
methods, products, instructions, or ideas contained in the material herein.

ISBN: 978-0-323-99124-7

For Information on all Academic Press publications visit our website at
https://www.elsevier.com/books-and-journals

Publisher: Nikki P. Levy
Acquisitions Editor: Anna Valutkevich
Editorial Project Manager: Howi M. De Ramos
Production Project Manager: Sajana Devasi P K
Cover Designer: Mark Rogers

Typeset by Aptara, New Delhi, India

Contents

Foreword

The book entitled "*Modern Intervention Tools for Rehabilitation*" by Dr. Dinesh Bhatia from North Eastern Hill University, Shillong, India, Dr. Meena Gupta and Dr. Prakash Kumar from Amity University Uttar Pradesh, Noida, India, is advanced, concise, and practical-oriented. All three are expertise in the field of rehabilitation and assistive technology. As we are living through swiftly changing external circumstances wherein innovation, experiential learning, and cross-collaboration play a pivotal role in the delivery of advanced rehabilitation tools for people with disability and the elderly population. The book would help those who are new to the field to acquaint themselves with the latest trends and help more experienced practising physicians, scientists, technologists, and researchers working in the field.

This book is a reference guide for readers who are a novice in the area or want to learn advanced rehabilitation techniques and medical insights related to new research or innovative ideas. It presents a comprehensive state-of-the-art approach to modern rehabilitation tools and practices. It provides a canvas to discuss emerging digital rehabilitation solutions, propelled by the ubiquitous availability of tools and devices, affordable, easy-to-use wearable sensors, and other technologies like virtual and augmented reality and mobile apps. The book's key features include coverage of advanced topics related to Brain Stimulation, Neuromodulation, Tele-rehabilitation, Transcranial Magnetic Stimulation, and Neurofeedback which are gaining momentum and being employed extensively nowadays. These technologies are seen as a solution to care access and providing quality healthcare capabilities at significantly lower costs, making them affordable to the patients and the service provider. It would help in enhancing the well-being and quality of life of the specially abled and elderly population who are dependent on such assistive devices. The book encompasses advanced topics in the field of Artificial Intelligence, MEMS, and Virtual/Augmented Reality with their applications.

This book is written for a wide audience with a common interest and believes it would immensely benefit clinicians, scholars, and researchers in Physical Therapy, Occupational Therapy, Healthcare Management, Biomedical Engineering, and Nano Technologies. It will help the reader make better guidelines for regulating digital technologies and allow the ethicist to appreciate the correct practices in rehabilitation. Overall, the book aims to help those who encounter enabling technologies as part of their research, scholarly work, or professional practice to better understand their potential, the psychosocial responses of the user, and how to use the technology to ensure faster and more effective rehabilitation and functioning. Finally, this book will help every kind of person that is curious as well to find out more about connected health.

All nine chapters help the reader to understand the present and shape the future of technologies in rehabilitation. The first two chapters summarise modern rehabilitation tools' concepts and ethical issues. Chapters third to eighth focus on the usefulness, effectiveness, and impact of Micro Electrical Mechanical Systems (MEMS) sensor, Brain Stimulation, Robotic Rehabilitation, Neural Prostheses, Virtual Reality, Augmented Reality, Machine Learning, and Artificial Intelligence. The book's final chapter addresses other advanced technologies, like the significance of mobile-based applications and smart homes in rehabilitation. Several illustrations of the different interventional tools and functional areas, complementing theoretical explanations, have been added to the book to understand the audience better.

Special mention to the Publisher Elsevier for their vision, encouragement, and assistance to the authors in making this book available to society in a limited time span. Credit must also go to their families, friends, and colleagues who would have supported the authors enormously in this noble endeavor.

Foreword by:
Prof. Vinod Kumar
Adviser and Former Vice Chancellor,
Jaypee University of Information Technology, Solan, India
Former Professor, IIT, Roorkee, India

Foreword

Rehabilitation is a set of assistive interventional tools and procedures designed to restore the optimum functioning of individuals with disability and/or medical conditions that dissuade them from doing daily life activities. The book *"Modern Intervention Tools for Rehabilitation"* has a special significance as it prepares for the new digital era of rehabilitation. The COVID-19 pandemic has uncovered the pressing need to prioritize health care and rehabilitation using advanced technologies that translate directly to the benefit of people's medical needs. Modern rehabilitation tools and techniques are to close the gap on sustainable development goal 3, for good health and well-being of humans and the society.

This book aims to provide a picture of the situation we live in today, in the age of modern information and communication technologies, it is possible to address them with modern tools and technologies. These technologies can enable efficient professional rehabilitation outcomes. The advanced technologies related to rehabilitation significantly help the healthcare industry with the adoption of new sensor technologies, robotic rehabilitation, virtual reality, assistive technologies, artificial intelligence, etc. for challenging and complex rehabilitation. Furthermore, this advanced technologies and their applications offer immense benefits in healthcare management in areas, such as autism, cerebral palsy, neurodevelopmental, and degenerative disorders including cancer, diabetes, cardiovascular, and trauma. These applications are also useful in the aftercare of patients and in assisting older adults in managing their activities of daily life.

It is most essential to have strong basic knowledge and understanding of advanced technological applications in health care before plunging into the field, whether you are a student or a professional. This book covers all the current problems, addresses research questions, and employs commercially available solutions. The authors discussed the technologies and their applications that are beneficial for clinics to community rehabilitation.

One of the many key reasons this book is so significant is its multidisciplinary authors. Dr. Meena Gupta and Dr. Prakash Kumar have been carrying out innovative techniques and information in rehabilitation over the past 15 years. Dr. Dinesh Bhatia has had a distinguished career in Biomedical Engineering and Advanced Healthcare Technology. They have long exposure, experience, and expertise in the conceptual and methodological problems besetting this field makes itself apparent in this book.

The real strength of this book is that it is richly studded with clinical usefulness drawn from a wide variety of set-ups and specialities which are very useful for students, researchers, and faculty/professionals of this discipline. These, along with the introductory frameworks, the potential for the present healthcare system, the extensive coverage of implementation, and the execution of advanced rehabilitation tools and techniques including related ethical issues, make this book a storehouse of the most comprehensive toolkit currently available.

I congratulate the authors led by Dr. Prakash Kumar for their excellent efforts in this noble endeavor. I am confident that the readers of this book will find it very educative and useful in their profession and clinical practice.

January 30, 2023.

Prof. Bhudev C. Das
Chairman & Hargobind Khorana Chair Professor
Dean, Health & Allied Sciences
Chairman, University Research Council (URC)
Vice President, Amity Science Technology & Innovation Foundation
(ASTIF), Amity University, Uttar Pradesh, India

Preface

We are living through a time of immense changes, where innovation, learning from experiences, and sharing ideas are essential to delivering the promises of advanced rehabilitation. This book provides a snapshot of the current state-of-the-art ideas around the world conducted by pioneers in the field. Our purpose is to highlight the modern tools for rehabilitation which have significantly evolved to help people with disability and improve their quality of life. We have tried to enumerate various facets of therapeutical interventions by sharing our knowledge and experiences with the readers. We leave this for our readers to judge.

The book would present many advanced types of research, innovation, and technologies that are evidence- and knowledge-based for the new learner. This book is a reference guide for readers who are a novice in the area or want to learn advanced rehabilitation techniques and medical insights related to new research or innovative ideas. The book presents a comprehensive state-of-the-art approach to modern rehabilitation tools and practices. It provides a canvas to discuss emerging digital rehabilitation solutions, propelled by the ubiquitous availability of tools and devices, affordable, easy-to-use wearable sensors, and other technologies like virtual and augmented reality and mobile apps. The book's key features include topics on Brain Stimulation, Neuromodulation, Tele-Rehabilitation, Transcranial Magnetic Stimulation, and Neurofeedback which are gaining momentum and being employed extensively nowadays. These technologies are seen as a solution to care access and better monitoring but not only this. It will provide patients with chronic diseases the possibility to manage their illness, thereby enhancing their well-being and quality of life.

This book is written for a wide audience with a common interest and believes it would immensely benefit clinicians, scholars, and researchers in Physical Therapy, Occupational Therapy, Healthcare Management, Biomedical Engineering, and Nano Technologies. It will help the legislator make better guidelines in regulating digital technologies and allow the ethicist to appreciate the correct practices in rehabilitation. Overall, the book aims to help those who encounter enabling technologies as part of their research, scholarly work, or professional practice to better understand their potential, the psychosocial responses of the user, and how to use the technology to ensure faster and more effective rehabilitation and functioning. Finally, this book will help every kind of person that is curious as well to find out more about connected health.

All nine chapters help to understand the present and shape the future of technologies in rehabilitation. The first two chapters summarize the concepts and ethical issues of modern rehabilitation tools. Chapters third to eighth focus on the usefulness, effectiveness, and impact of Micro Electrical Mechanical Systems (MEMS) sensor, Brain Stimulation, Robotic Rehabilitation, Neural Prosthesis, Virtual Reality, Augmented reality, Machine Learning, and Artificial Intelligence. The book's final chapter addresses other advanced technologies, like the significance of mobile-based applications and smart homes in rehabilitation. Several illustrations of the different interventional tools and functional areas, complementing theoretical explanations, have been added to the book to understand the audience better.

Finally, we would like to thank all people who have played a role in facilitating these ventures. We would like to thank the members of the Elsevier editorial and production team for their vision, encouragement, and assistance. Lastly, we would like to thank our families, friends, and colleagues, who have all helped us enormously in this endeavor.

Authors:
Dr. Meena Gupta, Amity University, Uttar Pradesh, India
Dr. Dinesh Bhatia, NEHU, Shillong, India
Dr. Prakash Kumar, Amity University, Uttar Pradesh, India

Acknowledgment

Writing a book involves enormous efforts from all collaborators and contributors. We would like to extend our sincere gratitude to all contributing authors for their painstaking efforts and support to incorporate finer details into the present book. The book entitled *"Modern Intervention Tools for Rehabilitation"* covers diverse topics that would enlighten the readers. We are grateful to our parents and family members for their kind support in allowing us to complete the book in time. We would like to thank Dr. Sonali Vyas, Department of Computer Science, UPES, Dehradun, Uttarakhand, India for her invaluable contributions in *Chapter 8, Machine learning, artificial intelligence technologies, and rehabilitation*. We acknowledge the constant support of our employers and Elsevier Publishers for allowing us to complete this challenging task of preparing the assignment in the available time. Without their continuous support and efforts, it would not have been possible to prepare and improve the quality of the book. We would like to thank Mr. Howell Angelo M. De Ramos, Editorial Project Manager, for his continuous and prompt help through every stage of book publication. A special thanks also goes to the authors, universities, and organizations who allowed them valuable time to pursue their research interests.

Finally, we thank Almighty God for his kind blessings, wisdom, and grace to enable us to complete this book. We hope the readers are benefitted from the contents present in the book.

Authors:
Dr. Meena Gupta, Amity University, Uttar Pradesh, India
Dr. Dinesh Bhatia, NEHU, Shillong, India
Dr. Prakash Kumar, Amity University, Uttar Pradesh, India

CHAPTER 1

Introduction

Concept of modern intervention tools

At the time of writing, we hope the devasting COVID-19 outbreak will finally end. International scientific communities consistently protect the public from the disease's clinical, economic, and political consequences. Because COVID-19 is a new disease, the impact on long-term outcomes in survivors is still emerging. A study of the sequelae in COVID-19 survivors' chronic illness is urgently needed. However, in some recent studies, post-pandemic, it is reported that the patients hospitalised with the disease had a new disability after discharge.

Even without a pandemic, the burden of disability is exponentially increasing in the present era. The steady increase in the burden of disease calculated using disability-adjusted life year (DALY) and the recent outbreak has focused on the high rates of illness, chronic pain, and reduced quality of life. When patient outcomes are below optimal, technological advancements rapidly emerge in our society to counter the gap. It directly or indirectly impacts individuals' and society's lives and behaviour. On the one hand, it enhances professionals learning abilities effectively, and on the other hand, it acts as a tool for clinicians and therapists to implement and execute services effectively. Overall, with the help of these advancements in rehabilitation, the client's quality of life has been meaningful and improved. The rehabilitation medicine and healthcare field is rapidly expanding through technological innovation advancements. As a result, many different areas of human health diagnostics, treatment and care are emerging.

Concept and history of rehabilitation

Rehabilitation is a program or process for returning an individual to health after an injury, illness, or addiction. Rehabilitation usually takes the form of therapy sessions. In these sessions, patients perform specific exercises for a fixed period under the supervision of a health professional, typically a physiotherapist or an occupational therapist. The overall goal of rehabilitation is primarily to recuperate a patient from impairment or disability and improve mobility, functional ability and quality of life. This impairment can result from ageing, injury, neurological disease, or neurodevelopmental

Modern Intervention Tools for Rehabilitation.
DOI: https://doi.org/10.1016/B978-0-323-99124-7.00007-9

disorders. Whether these injuries occur in isolation or in combination, they require an interdisciplinary team of specialists and a well-coordinated, integrated, holistic rehabilitation plan. These efforts also help to maximise functional independence; promote patients' highest quality of life, successful reintegration, and active participation in their families and society.

Philosophy of rehabilitation grew significantly during the polio epidemic during the early 1900s and gained international attention when president Franklin Delano Roosevelt, after suffering from lower limb paralysis due to polio, spent time learning how to walk at a unique facility in Warm Springs, Georgia. His-personal success inspired him to purchase and expand the facility, which is believed to be the first such place in the country to specifically provide rehabilitative care. Their return home with limb loss, paralysis, and other physical impairments during World War II necessitated more comprehensive rehabilitative care. In response, the US military established a training program in 1942, which was headed by Dr Howard Rusk (considered the "father of comprehensive rehabilitative care").

Rehabilitative medicine takes a team approach that starts during the acute phases of combat casualty care, as previously described. Inpatient teams most frequently include physiatry, physical, occupational, and recreational therapists; dietitians; social workers; rehabilitation counsellors; and speech-language pathologists, case managers, prosthetists, and orthotists. After discharge, patient care continues with outpatient physical and occupational therapy, driving rehabilitation, recreational/motivational therapy, and music/art therapy. The importance of long-term rehabilitation and the involvement of multi-disciplinary professionals make it practical and financial constraints. Further, due to limited coverage in health care plans, many individuals receive limited or no support to continue rehabilitation or maintain the progress they made in treatment. Technology-based rehabilitation has provided a potential solution that can supplement traditional rehabilitation methods as per the individuals' convenience.

Technological advancements come into existence through the past decades' major health reforms, which should not be overlooked. In the present era, technology can potentially transform the health experience for the better. Incentives need to shift from older financial models that reward hospitals and clinics for expensive procedures and tests rather than for keeping their patients healthier for longer. There are two main types of technological equipment that rehabilitation providers can use: that which has been designed for the general population and that which has been specifically designed for people with "special needs." Technology designed for use by the general population has high face validity for clients because of its normal use. It would include home computers, the Internet, palmtops, mobile telephones and, more recently, the potentially helpful global navigational hardware. The use of technology reduces patient hospitalisation times and costs and increases the number of patients who can be treated at the same time. Another positive aspect of modern rehabilitative models is direct and continuous interaction between the patient and health care provider, which increases compliance to treatment and patient safety during rehabilitation care.

Benefits of modern intervention tools in rehabilitation

The main incentives for the use of modern or advanced intervention tools in rehabilitation can be summarised as their ability to reduce the physical and cognitive workload of the therapist and other rehabilitation professionals; the possibility to store log relevant data for more objective diagnosis, prognosis and also help in to develop new innovative intervention; their potential in making rehabilitation process more accuracy and engaging experience. The following important things are highlighted:

- Travelling: Patients must travel to rehabilitation clinics or medical centres to receive treatment. This is sometimes inconvenient, time-consuming, and challenging for an individual with physical disabilities. Modern rehabilitation with technology such as robots can allow patients to receive treatment in more convenient locations such as at home or at work. The benefit we expect from this modern approach is twofold. Most patients feel psychologically better in their environment than in the hospital and rehabilitation speed is improved. In addition, we aim to exploit patients' increased motivation when exercising with a tool similar to a gaming console. The ability to do high-quality exercises without being monitored by a therapist helps to reduce healthcare costs through the minimisation of expensive human contact hours. Recently, gaming consoles have gained much attention when used in rehabilitation.
- Availability: Rehabilitation exercises involve manual hands-on treatment with the physiotherapist or occupational therapist. This is tiring for both the patient and the therapist and is challenging to perform continuously for extended periods in cases where frequent treatment is advisable over long rehabilitation sessions. Rehabilitation with technology is often not affected by fatigue and can operate for an unlimited time.
- Subjectivity: Therapists evaluate patient disability and recovery based on their own opinion. This can be inaccurate and lead to treatment that

may not be optimal for the patient's condition. Advanced rehabilitation through technology provides an accurate, objective measure of a patient's disability characteristics at specific joints and muscle groups in the body. In one way, it allows researchers or clinicians to deliver rehabilitation the exact same way to all individuals, i.e., standardising the delivery of rehabilitation. In other truth, rehabilitation may not be best provided in a one-size-fits manner, even if the item sets are tailored. With the help of technology, more individualised rehabilitation in terms of tailoring the item sets trained for each patient.

- Motivation: In conventional rehabilitation, methods tend to involve repetitive movement and will likely cause a reduction in the voluntary effort put in by patients. Modern rehabilitation methods where therapy sessions is conjunction with virtual games to give the patient fun and engagement while performing rehabilitation.
- Complex and challenging rehabilitation: Conventional rehabilitation often provides basic treatment with available tools and equipment. It isn't easy to provide multiple tasks at one time, like manipulating multiple joints simultaneously. However, with the help of advanced technology, these situations can be managed and capable of generating complex movements.

Though there are many different and effective rehabilitative technologies, here are some of the most innovative and in-use rehabilitation technologies:

IT in rehabilitation

The application of information and communication technologies across the entire range of functions involved in healthcare practice and delivery is generally defined as e-health. This could include various information such as patient's medical records (EMR), billing and payment information, and employee and hospital information. One of the implementations of E-health today involves the use of the Internet for storing, accessing, and modifying healthcare information. The major thrusts behind the introduction and deployment of e-health are to obtain efficiencies in healthcare delivery and management, improvement in healthcare quality, cost reduction, reduction in medical errors, and moving healthcare resources to the place of need.

Several challenges and obstacles must be addressed before e-health becomes prevalent. These challenges primarily deal with how medical and healthcare information has been collected and stored, the lack of technologies and the potential cost of digitising the existing processes and tasks. Even as some of the information is being digitised and converted to the electronic domain, it is observed that several different systems of e-health are evolving. Some of these are closed or non-standardised proprietary systems, which do not support interoperability with other systems, thus resulting in islands of e-health that are difficult to interconnect

Over the years, interactive computer-based systems have provided crucial support to clinics, hospitals and other rehabilitation centres. These systems have continued to influence the manner in which clinical tasks are organised and fulfilled in terms of performing tests, diagnosis procedures, and treatment methods, as well as storing, analysing and accessing patient and staff information. Currently, the computer-based systems used in healthcare settings of high standards are the result of joint efforts of clinicians, software developers and clinical informaticians, hence triggering the outcome of the desired system to outdo existing applications.[1] Healthcare systems, whether in the form of desktop applications or mobile applications, have managed to replace paper-based systems to a large extent. One of the major themes of interest for biomedical IT systems in today's time comprises web-based and wireless healthcare facilities. Reasons include broad access vicinity and quick and easy access to information.

Robotics therapy

Robotic systems allow clinicians to increase treatment duration, intensity and specificity compared to traditional physical therapy. The advances in robotics in rehabilitation medicine are impressive and can improve the therapy of those suffering injury and recovery. Rehabilitation robots assist physical therapists with patients to improve recoveries in neurological impairments, like those recovering from a stroke or a brain trauma. Compared to manual therapy, rehabilitation robots have the potential to provide intensive rehabilitation consistently for a longer duration. They are not affected by the skills and fatigue level of the therapist. These devices help the patient's level of independence and quality of life and assist clinicians or therapists in better understanding a patient's recovery and progress. Consequently, individuals who undergo therapy using robotic systems could do better than patients who receive traditional physical therapy. Robotic rehabilitation is particularly appealing in patients with conditions such as spinal cord injury, stroke, and traumatic brain injury. A major goal of therapy for these patients is to achieve motor recovery. Increasing the duration, intensity and specificity of therapy – as it is possible by leveraging robotic systems for rehabilitation – is in tune with this goal.

Since robots are well suited for repetitive tasks and can be designed to have adequate force capabilities, their use in the execution of these exercises will be able to reduce the physical workload of therapists. It can potentially allow the therapists to simultaneously oversee the treatment of multiple patients in a supervisory. By using robotic devices, diagnosis and prognosis can be made more objectively with the help of quantitative data, and comparisons between different cases can also be made more quickly. Several successful rehabilitation robots have undergone clinical trials and are currently being used in hospitals and clinics for neuromotor rehabilitation. However, the research and development of advanced robotics for medical rehabilitation are still early, and further research and development in this area are becoming increasingly urgent. Robots can treat patients without the presence of a therapist, enabling more frequent treatment and potentially reducing costs in the long term. In addition, a rehabilitation robot can measure quantitative data to accurately evaluate the patient's condition. Using specially designed virtual games with the robot can provide an entertaining therapy experience, encouraging the patient to put in their effort into the exercises.

Robot-guided therapy can either use assistive or resistive methods, and it is not currently clear which is more effective. In active assistance training, a therapist or robot assists the patient through the desired motion. The benefits of active assistance include stretching the muscles and connective tissues, reinforcing a regular pattern of movement, and allowing the patient to practice more complex tasks.[2] Active assistance also increases the intensity of training, since, with assistance, more motions may be completed in less time.

Resistive training methods work to facilitate rehabilitation by making task completion more difficult during training by applying forces that resist or perturb the motion. Individuals moving in a force field that perturbs their motion will adapt to generate forces that counteract the field, resulting in a regular motion within the field. The adaptation will persist for a short time after the field is removed. This after-effect has led to error enhancement training, in which an individual's errors during a motion are exaggerated. Once the disturbance is removed, the after-effect results in a more correct movement; however, the corrected motion typically only persists for a short amount of time.

The robot-assisted exercise can be summarised into three broad scenarios: haptic simulation, challenge-based, and assistive.

Hepatic simulation: Robots are used for haptic rendering in virtual environments—in the virtual environment, the subject can exercise with a variety of interaction tasks, generally inspired by activities of daily living

(ADLs). Robots, combined with visual displays, allow joint visual and haptic interaction with virtual objects.

Challenge-based: The robot provides disturbances and/or perturbations to make a task more difficult or challenging with respect to performance without the robot. Several approaches have been proposed. During exercise, the robot can generate perturbations that oppose the subject's movement or compel the subject to provide a greater force. The robot may also be programmed to generate dynamic environments that have a destabilising effect, for example, negative viscous forces. The rationale underlying challenge-based scenarios is that making the task more difficult during training will later result to an improved performance in unassisted or unperturbed exercises.

Assistive: The robot provides forces that facilitate task performance or task completion. The goal is to help the subject move the impaired limb in specific goal-oriented tasks such as grasping, reaching, and walking.

Home-based rehabilitation with the help of robots has been shown to help maintain an individual's ability to perform the activity of daily living. Several methods have been adapted for home use but are limited to individuals with mild impairments. Ideally, moderately impaired individuals would also be able to benefit from home-based rehabilitation. For example, the SMART system[3] incorporates a motion tracking system to monitor the performance of daily tasks and rehabilitation exercises, an online database that allows therapists to monitor patient performance, and a visual feedback system that therapists may use to provide instruction. These home-based methods, however, use a home computer with limited accessories that can only provide limited assistance forces and have a limited workspace. These methods can provide some benefits, but rehabilitation is limited to people with relatively high motor function. The challenge is to develop safe and affordable rehabilitation for individuals with moderate impairment. Bimanual rehabilitation is ideal for home-based stroke therapy since much of the required force could be provided by the person's sound limb or by using robots. Robotic bimanual rehabilitation uses a robotic device to assist an individual in making bimanual motions.

Robotic system for gait rehabilitation

Several lower extremities rehabilitation devices have been developed in the last decade for gait training during walking. Robots have the potential to provide accessible, precise, and physiological gait patterns adapted and reproduced with higher accuracy, making it possible to test and optimise the biomechanical gait pattern (speed, step length, amplitude) to get an optimal effect. In addition, patient progress can be monitored, and training sessions can be longer without the force limitations and inconsistency of conventional human-regulated therapy, relieving strain on therapists. Robots developed for lower limb rehabilitation can be classified based on their mechanical structure as (a) end-effector devices and (b) as exoskeletons. While most of these systems represent unique engineering designs and offer many new possibilities for gait training after stroke, they also have important limitations that do not allow their widespread use in clinical practice. This high cost of existing robotic systems for gait rehabilitation is a huge hurdle that needs to be overcome so that robotic systems become widely accepted in clinical facilities. An end-effector lower limb rehabilitation robotic device has indirect control of the patient's limbs by applying forces on the patient's body using a structure that includes the system's actuators outside the patient's body. An exoskeleton is a mechanical structure that includes the actuators worn by an operator. Anthropomorphic exoskeletons attempt to mimic the kinematic structure of the human skeleton. As they work in parallel with the user's limbs, mechanical limits can be implemented directly, and the risk of collisions is eliminated. The interactive participation of the patient should also be considered in gait rehabilitation because it can accelerate the rehabilitation process, emphasising the importance of assistive robots. For example, Robotic Gait Rehabilitation (RGR) Trainer was developed to target secondary gait deviations (i.e. gait abnormalities that result from compensatory movements associated with a primary gait abnormality in patients post-stroke).

Virtual reality

Virtual Reality (VR) is an emerging technology that holds tremendous opportunities for developing practical assessment and multimodal treatment techniques in highly controllable environments. VR has recently drawn professionals' and patients' attention in several fields, including psychology, physical and neurological rehabilitation. VR, just like name suggests, is a virtual world of computer-generated elements completely isolated from the real world where patients can perform fun activities or games synonymous with their required routines. In recent years, virtual reality has grown immensely and established as a novel adjunctive therapy.[4] Through its capacity to allow the creation and control of dynamic 3D, ecologically valid stimulus environments within which behavioural responses can be recorded and measured, and these options are not available with traditional methods.[5] For example, it is effective in pain management because it distracts the brain from thinking about pain. In the research, Turolla[6] evaluated the effectiveness of non-immersive VR treatment for restoring the upper limb motor function and its impact on the activities of daily living in post-stroke patients. VR rehabilitation in post-stroke patients seems more effective than traditional interventions in restoring upper limb motor impairments and motor-related functional abilities. With VR, instead of mindlessly repeating an exercise, the same movements can be done within a virtual environment where the patient is engaged in an immersive and gamified experience.

VR systems have three formats: non-immersive, semi-immersive and fully immersive. The main concept which is frequently used is "immersion" with VR. "Immersion" refers to the sense of being involved in a task environment without considering the time and real world and up to which extent high fidelity necessary inputs (e.g. sound waves, light samples) are supplied to diverse sensory modalities (touch, audition, vision) to build a powerful illusion of realism. A non-immersive VR system utilises the usual graphics terminal with a monitor, typically a desktop system, to view the VR environment using some portal or window. This format imitates a three dimensional graphics environment on a television or flat panel within which the user can interact and navigate. A semi-immersive VR system is a relatively new implementation comprising a comparatively high performance graphics computing system and an outsized projection plane to display scenes. A fully immersive system gives a sense of presence. Still, the level of immersion depends on various factors like the field of view of resolution, contrast, update rate and the illumination of the display. Generally, an immersive VR system clubs a computer, body tracking sensors, a specialised interactive interface such as a head-mounted display or an outsized projection screen encasing the user (e.g. CAVE–Cave Automatic VEs where VE is projected on a concave surface) and real-time graphics to immerse the participant in a computer-generated world of simulation to perform alterations naturally with the body and head motion. Thus, this format leads us to adopt an immersive learning environment for health care services and applications presently and in the future.

Augmented reality

Generating new images from digital information in the real physical environment of a person, simulating an environment where the artificial and real are mixed. AR is a subset of VR (not a complete virtual reality). It must be differentiated from virtual reality (VR), in which additional data such as sound, text, or video are introduced, giving rise to multimedia virtual environments to enhance the user's learning experience. AR is derived from VR but blends these virtual environments with real ones, enhancing the interaction with real life.[7] In rehabilitation, it has been developed mainly for motor and cognitive rehabilitation, which is considered a new intervention method. AR can be used as a working tool to complement the treatment conducted by the physiotherapist, as it generates safe environments that are similar to the patient's natural environment.[8] Rehabilitation using AR has shown better results than repetitive movements practised alone as AR allows the better orientation of the exercises toward objectives with greater patient motivation and is enjoyable to use.[9] They provide new experiences to patients during rehabilitation sessions, increasing engagement and improving physical outcomes[10] and create exciting opportunities to provide low-cost physiotherapy at home.[11]

Since the advent and extreme usage of smart phones in recent times, most of AR applications are based on this new invention. Hence focusing the smart phones, there are two major AR formats. According to Pence, 2010[12]:

a) Marked or mark based AR system utilises a two-dimensional barcode normally QR code (quick response code), to connect a mobile phone and/or personal computer for overlaying information digitally on real-world objects or usually on a website;

b) Mark less AR system employs location-based services like GPS (Global Positioning System) used by cell phones to serve as a platform for adding native information to a camera vision.

Transcranial stimulation

Transcranial stimulation techniques are in use today: Transcranial Magnetic Stimulation (TMS) and transcranial Direct Current Stimulation (tDCS). TMS elicits excitability changes in the underlying cortex via the production of large, rapidly modulating magnetic pulses within a coil placed on the scalp to induce a current in the underlying tissue. tDCS involves passing minimal currents in the order of a few milliamps through the skull via two large electrodes placed on the scalp. Although both techniques have a significant potential role in clinical studies, tDCS has several practical advantages as a prospective rehabilitation tool; it is well tolerated, relatively easy to administer and portable. In addition, although the neurophysiological effects of tDCS and TMS are similar, their mechanisms of action are at a cortical level. Therefore their potential impact on patients are likely to be distinct. Both these approaches have been shown to improve motor function in a clinically relevant task in the affected hand.[11,13] A recent study also demonstrated that tDCS can shift the inter-hemispheric balance of motor-related activation in healthy controls[14] suggesting this "rebalancing" as a putative mechanism for observed behavioural improvements. However, several questions to be answered before transcranial stimulation paradigms can be used as adjuncts to rehabilitation in clinical practice: in particular, whether the duration of effects of stimulation can be increased; whether this model of inter-hemispheric imbalance holds true for the wider clinical population and whether the magnitude of the impact would be improved if stimulation was applied in the acute or sub-acute period.

Android-based system for rehabilitation

It is a form of telerehabilitation that uses advanced telecommunication technologies to exchange health information and provide healthcare services across geographic, time, social, and cultural barriers. It can be defined as "remote rehabilitation care" by which patients can be examined, analysed, monitored and rehabilitate without being co-located with the therapist and clinician. Two types of interaction occur in Telerehab- the Real-time (synchronous) and the Store and Forward (asynchronous). It has been recognised in the healthcare industry that long-term continuous monitoring is critical to caring for people with chronic conditions such as non-communicable diseases. This avoids hospitalisation and enables those in geographically isolated settings to access specialised and preventive medicine. There are many language based platforms to develop mobile phone software like the proposed android-based application. Different mobile phone brands use different programming platforms. Nowdays, mobile phones are standard tools for society, and because of this, exercise can be performed almost everywhere and at any time. This type of intervention is seen as an enabler of change worldwide because of its high reach and low-cost solutions. It enables access to care for individuals in remote areas or those with mobility issues associated with physical impairment, access to transport and socioeconomic factors. In addition, it cuts down the associated travel costs and time spent travelling for both the healthcare provider and the patient. Research has found that the rehabilitation needs for individuals with long-term conditions such as stroke, TBI and other neurological disorders are often unmet in the patient's local community. This application introduced live interactive video conferencing that allows for consultations, diagnostic assessments and delivery of treatment interventions and provides verbal and visual interaction between participants. It also allowed therapists to create personalised physical exercises, remotely customisable to patients' needs, assign exercises prescription over a period of time, remotely monitor patients' vital signs, video-call or chat with patients in real-time and assess the execution of the exercises.

References

1. Pagliari C. Design and evaluation in eHealth: challenges and implications for an interdisciplinary field. *J Med Internet Res.* 2007;9(2):e15.
2. Hummel F, Celnik P, Giraux P, et al. Effects of non-invasive cortical stimulation on skilled motor function in chronic stroke. *Brain.* 2005;128:490–499.
3. Zheng H, Davies R, Zhou H, et al. Smart project: application of emerging information and communication technology to homebased rehabilitation for stroke patients. *Int J Disabil Human Dev Spec Issue Adv Virtual Real Ther Rehabil.* 2006;5(3):271–276.
4. Saposnik G, Levin M. Virtual reality in stroke rehabilitation a meta-analysis & implications for clinicians. *Stroke.* 2011;42(5):1380–1386. doi:10.1161/STROKEAHA.110.605451.
5. Tsirlin I, Dupierrix E, Chokron S, Coquillart S, Ohlmann T. Uses of virtual reality for diagnosis, rehabilitation and study of unilateral spatial neglect: review and analysis. *Cyberpsychol Behav.* 2009;12(2):175–181.
6. Turolla A, Dam M, Ventura L, et al. Virtual reality for the rehabilitation of the upper limb motor function after stroke: a prospective controlled trial. *J Neuroeng Rehabil.* 2013;10:85.
7. Portalés C, Lerma JL, Navarro S. Augmented reality and photogrammetry: a synergy to visualize physical and virtual city environments. *ISPRS J Photogrammetry Remote Sensing.* 2010 Jan;65(1):134–142. doi:10.1016/j.isprsjprs.2009.10.001.
8. Manuel N, Navarrete M. La realidad virtual como arma terapéutica en rehabilitación. *Rehabil Integral.* 2010 Jun;5(1):40–45.
9. Sveistrup H. Motor rehabilitation using virtual reality. *J Neuroeng Rehabil.* 2004 Dec 10;1(1):10. doi:10.1186/1743-0003-1-10.
10. Postolache O, Monge J, Alexandre R, Geman O, Jin Y, Postolache G. *Advanced Systems for Biomedical Applications.* Cham: Springer; 2021 Virtual reality and augmented reality technologies for smart physical rehabilitation.
11. Cary F, Postolache O, Girao PS. Kinect based system and serious game motivating approach for physiotherapy assessment and remote session monitoring. *Int J Smart Sensing Intell Syst.* 2020 Jan;7(5):1–6. doi:10.21307/ijssis-2019-131.
12. Pence HE. Smartphones, smart objects, and augmented reality. *Ref Libr.* 2010;52(1–2):136–145.
13. Fregni F, Boggio P, Mansur C, et al. Transcranial direct current stimulation of the unaffected hemisphere in stroke patients. *Neuroreport.* 2005;16:1551–1555.
14. Stagg C, O'Shea J, Kincses T, Woolrich M, Matthews P, Johansen-Berg H. Modulation of movement-associated cortical activation by transcranial direct current stimulation. *Eur J Neurosci.* 2009;30:1412–1423.

CHAPTER 2

Ethics in modern rehabilitation system

Every people living in the community has freedom, interest, and rights of others. Advanced technological development in the healthcare system enabled the public to carry out various health-related activities via the Internet. This move towards e-healthcare envisaged reducing the cost of healthcare provision, improving the quality of care and reducing medical errors. In addition, it gives the public the information they need to make informed choices about their health and healthcare. It also helps to develop the capacity to process large volumes of data for public health surveillance quickly and efficiently to allow early detection of threats (e.g. flu outbreaks, adverse drug reactions). Creation, modelling, management and sharing of health data and knowledge to support data analysis and timely decision-making in medicine and healthcare created various opportunities and challenges for healthcare providers and patients. In the process, all relevant information about the patient is stored in the computer system in the hospital record section. However, keeping such information in the computer or server-based record section is not at all secure due to the availability of technology that may circumvent or get access to the said information. This could threaten (confidentiality) privacy, data integrity, and patient security. Therefore, we need the proper law and regulations that address the above concern and avoid unnecessary conflicts. Such rules can be written as part of legal norms. As we are about to explore, however, legal norms alone are not enough: for a coherent and active society, we also need ethics. Ethics is a constructed norms of internal consistency regarding what is right and wrong. The four principles of Beauchamp and Childress medical ethics are also considered in a modern rehabilitation setting.

- Autonomy: Patients have autonomy of thought, intention, and action when making decisions regarding health care procedures. Therefore, the decision-making process must be free of coercion or coaxing. In order for a patient to make a fully informed decision, she/he must understand all risks and benefits of the procedure and the likelihood of success.

Modern Intervention Tools for Rehabilitation.
DOI: https://doi.org/10.1016/B978-0-323-99124-7.00008-0

- Beneficence: Requires that the procedure be provided with the intent of doing good for the patient involved. Healthcare providers must develop and maintain skills and knowledge, continually update training, consider the individual circumstances of all patients, and strive for net benefit.
- Non-maleficence: Requires that a procedure does not harm the patient involved or others in society. The inclusion of technology in modern rehabilitation sometimes makes it problematic for clinicians and professionals to successfully apply the do not harm principle.
- Justice: The idea that the burdens and benefits of new or experimental treatments must be distributed equally among all groups in society. When evaluating justice, the health care provider must consider four main areas: fair distribution of scarce resources, competing needs, rights and obligations, and potential conflicts with established legislation.

Concept of law and ethics in healthcare

Law

Laws are societal rules or regulations that are advisable or obligatory to observe. Failure to observe the law is punishable by the government. Laws protect the welfare and safety of society and resolve conflicts in an orderly and nonviolent manner. When an individual violates a specific legal norm, clearly defined procedures for establishing guilt and prescribed sanctions are in place. Constitutions, laws, regulations, and other similar documents provide us with precise guidelines on how to act as individuals in a society

Ethics

Ethics is a set of moral standards and a code for behaviour that govern an individual's interactions with other individuals and within society. 'Morality' is what people do, in fact, believe to be right and good, while 'ethics' is a critical reflection about morality and the rational analysis of it. For example, "Should I terminate the pregnancy?" is a moral question, whereas "How should I go about deciding?" is an ethical concern. In other words, ethics implies a rational and systematic study of moral issues. In health care, the study of ethics has helped to develop principles and decision-making approaches that guide clinical practice. It has also helped clarify how the health professions can best fulfil their responsibilities in serving patients, communities, and society. In healthcare practice, the duty of care to protect life and health is superseded by the duty to respect autonomy. This means competent patients have the right to refuse any medical intervention.

Thus, it is both legally and professionally unacceptable to force treatment on competent patients because the doctor thinks it is in their best interests.

Laws are mandatory rules to which all citizens must adhere or risk civil or criminal liability. Ethics often relate to morals and set forth universal goals that we try to meet. However, there is no temporal penalty for failing to meet the goals as there is apt to be in law. The connection between medical law and medical ethics is revealing. It might be thought that the two would be closely connected. The law sets down minimally acceptable standards, while ethical approaches may include deciding what would be the ideal way for a person to behave.

Informed consent

Autonomy refers to one's moral right to decide one's course of action. Thus, if the ethical principle of autonomy is applied to healthcare practice, it means that clients must be given sufficient information about healthcare before the intervention or other clinical procedure and then permitted to decide for themselves about the proposed treatment. For example, patients must provide their informed consent before performing a gastrointestinal endoscopy, a procedure carrying a considerable risk of harm. This involves presenting patients with the factual details, the advantages and the disadvantages of undertaking such a procedure, and the potential complications of this proposed intervention as the basis for making an informed choice. Nevertheless, informed consent should protect patients by providing complete information to make an informed decision. If the healthcare professional in performing a procedure mistakenly believes that the patient's consent has been gained when, in fact, it has not, the healthcare professional is at risk of litigation, together with those individuals who acted as double-checkers in the situation.

Ethical issues in modern rehabilitation industry

Regulatory compliance and liability issues are generally written to address existing technologies but cannot always foresee future ones. That is because the progression of the e-health industry is dependent upon and influenced by the rapidly changing advances in the law. Compounding the lag between law and technology, it can take considerable time to update the law. Transportation laws written for railroads could not foresee automobiles or commercial aviation. Just like the same medical ethics could not have contemplated modern rehabilitation services. The modern rehabilitation industry includes mobile applications, wearables, implantables, and related technologies, which are not much explained in existing law, regulations and ethical standards. Major issues in adopting these advanced technologies lie in patient privacy, trust, product liability and negligence in handling medical data. The implementation and adoption of e-health are lacking in many countries, including the USA and European countries, due to barriers such as legal protection on privacy and other liability issues. Many situations arise in digital rehabilitation practice that requires a moral decision. For example, should a parent have a right to refuse immunisation for their child? Is primary health care a right or a privilege? Who dictates client care—the physician, the attorney, or the medical insurance carrier? Should children with severe congenital disabilities be kept alive? Should a woman be allowed abortion for any reason? Should everyone receive equal treatment in medical care? Should people suffering from a genetic disease be allowed to have children? Should individuals be allowed to die without measures being taken to prolong life? Should criteria be developed to determine who receives donor organs? Should stem cell research be limited? None of these questions has an easy answer. However, one may sometimes have to make these decisions or be in a position to assist those who make such decisions. We do not attempt in this book to determine right or wrong for the ethical issues in modern medicine. The purpose is to present the law and the pertinent facts in the ambulatory health care setting and raise some questions for consideration. As health professionals, we must live and act to respect ourselves and others and encourage others to have respect for themselves. We need to know what we are to become and how we can become better than we are. Even when our opinions differ from the clients we serve, clients always deserve respect and dignity from us.

For the digital advanced modern rehabilitation, ethical standards and laws designed to protect clients and establish guidelines for the professionals represent efforts to create a climate for an equitable exchange between client and provider. Every institution providing modern rehabilitation through advanced technologies should establish an ethics committee and update it accordingly. The unit that is in charge of executing the program should be aware of the clients' rights and safety. They have a system that decides the appropriate course of action in the case of conflicting opinions and efforts to create a climate for an equitable exchange between client and provider. Although advanced rehabilitation may enhance quality health care, it demands greater coordination for a client to benefit from it, increasing the cost of care.

Another critical aspect of medical law and ethics is that professionals are of high standards. They should be transparent, open, and knowledgeable about their personal choices and belief and be able to recognise vast diversity in a pluralistic society. Rehabilitation professionals must feel comfortable in a "servant" role while maintaining their own integrity and the respect of their clients. Overall, health professionals' first mandate is to "not harm"; the use of technology brings many complicating issues into play. This telemedicine presents unique ethical problems, such as maintaining the confidentiality of the information and the privacy of patients and safeguarding the integrity of information systems.[1] Remote patients could benefit or be disadvantaged by virtual care (e.g., lack of access to the Internet, smartphones, or other technology should not prevent children from accessing their medical system).[2] Indeed, the principle of justice includes equal access to care and fair distribution of technology for marginalised communities.[3] Ideally, the most significant advantage for patients should be equitable and quick access to healthcare through telemedicine services. However, this aspect is still controversial and, in some cases, has been exacerbated during the COVID-19 pandemic (e.g., inequitable access to care, unsustainable costs in a fee-for-service system, and a lack of quality metrics for novel care-delivery modalities). Therefore, the practice of telemedicine needs significant improvement, with specific rules and codes of conduct to be correctly put into practice for a sustainable program to be built.

Key ethical issues in the clinical delivery of telehealth vary by profession. Generally, they include avoiding harm, defining the professional relationship, obtaining informed consent, discussing the limits of confidentiality, maintaining patient confidentiality, operating within one's boundaries of competence, establishing fees and financial arrangements, avoiding false or deceptive statements, providing services to patients served by other professionals, providing consultations; avoiding sexual intimacies with patients, maintaining records and data, and avoiding use of confidential material for didactic or other purposes. Although most professions have delineated appropriate definitions of professional conduct for face-to-face treatment, the applicability of these definitions can easily become blurred when using technology. For example, most healthcare professionals are obligated to ensure the confidentiality of patient identity and records. In some professions, practitioners and researchers must also inform the patient or subject of the limits to privacy. Yet when using electronic tools, especially the Internet, professionals can easily be dazzled by the excitement of instantaneous communication with a patient halfway around the globe and easily forget their ethical requirements.

Similarly, professionals may not be thinking of their patient records when having an office computer serviced at a neighbourhood computer repair shop. Nonetheless, they may expose their patient files to unauthorised access and reproduction. Another area in which professionals might err in violating patient confidentiality is using email support groups for research without obtaining group members' consent.

Privacy and security

Privacy as a concept is neither clearly nor easily defined, and yet the fear of losing personal privacy is related to information and communication technology. Medical records are the lifetime history of every patient. Thus proper recording, keeping and retrieving remain burdensome but necessary for accurate medical assessment and clinical interventions. Specific concerns relate to the amount of personal information gathered, the speed it is transported (raising concerns about its accuracy), the duration of time that personal information is stored and the kind of information to be transferred. Tavani (2007)[5] summaries three views of privacy as

- Accessibility privacy is physically being left alone or being free from intrusion into your physical space.
- Decisional privacy relates to the freedom to make personal choices and decisions.
- Informational privacy concerns the control over the flow of personal information, including the transfer and exchange of information.

To protect the privacy of the user of advanced technology, the aim of technology and the way it is used by service providers or care organisations, as well as how personal data will be processed electronically, must be described explicitly in a privacy statement and communicated to the user. The privacy statement should include the name and function of the person who has final responsibility for the daily processing of the personal data; the location where the data are stored in paper and/or electronic form. It should also explain the specific aim, content and usage of the data and the person(s) who informs the user about this, the person(s) that can be contacted if personal data prove to be incorrect. In case of any error, what measures are taken to prevent inspection, mutation or data removal by unauthorised persons? Personal data should be processed only after the explicit informed consent of the user. Access to confidential information and sharing it with other professionals should be well controlled. Additional measures are to store personal data in devices only if necessary for follow-up care and for as long as necessary, then deleting them or transferring them to a secure central location. Another feature is that cookie also invades e-health system users' privacy. The cookies collect the user's web browsing information and store the information on the user's hard drive. The data will be sent automatically or retrieved by the websites to track their online activities to build a market profile of computer users. This way of collecting information from Internet user using cookies violates the law. They may manipulate and transfer the potential information to other companies or institutions that misuse it unfairly or for purposes other than those consented to by the individuals. Therefore, this industry needs to provide appropriate technical and organisational measures for privacy protection.

Technology system security view as a process, not a product of the technology. It is related to confidentiality (protecting against unauthorised disclosure of information to third parties), integrity (preventing unauthorised modification of data and files), and availability (preventing unauthorised withholding of information from those who need it when they need it). Complete security can be achieved if mechanisms for data security, system security and network security are available. A security policy should be created that ensures that ethics and privacy issues are respected and that information security is not breached.

Jurisdiction of courts

As the community of technology users for e-health purposes grows increasingly diverse and the range of Internet healthcare interaction expands, disputes of every kind may be expected to occur. Technology commitment will be breached; negligence and privacy invasion may occur. Although many of these disputes will be settled informally, others may require formal mechanisms. When the patient concerned wants to bring a proper legal action, questions like who has jurisdiction to hear the case? How would a judgement be enforced against a defendant in another country? This jurisdiction issues in the context of online e-health activities present one of the most significant challenges to the current system and how to address disputes involving two or more countries. The current laws regulating jurisdictional issues that apply geographical criteria. Accordingly, the courts will not expand their jurisdiction beyond its physical borders. However, with e-activities, geographical boundaries become meaningless. Exercise of jurisdiction over a person beyond geographical boundaries becomes a primary factor to be considered because of the borderless nature of e-activities.

Exemplary e-health organizations

In an attempt at self-regulation, various Internet-based groups have evolved to help set standards and guidelines for various types of health care Websites on the Internet. Though not exhaustive, the list that follows will help direct you to organisations focused on general health care Websites rather than speciality professional Websites.

The Internet Healthcare Coalition (IHC). The IHC (2000) has developed an e-health code of ethics "to ensure that all people worldwide can confidently, and without risk, realise the full benefits of the Internet to improve their health." The code of ethics outlines many guiding principles to help practitioners adhere to standards for e-health. For information, go to (www.ihc.net/).

Health Internet Ethics (Hi-Ethics). In 1999, several major Internet health care companies came together to discuss ethical, credibility, and integrity issues for consumer health sites. Among the topics were commerce regulations, privacy issues, content quality, and responsible advertising practices. The Hi-Ethics principles give protection to both users and providers of health Web sites. Users benefit because the principles provide strong and verifiable rules to prevent privacy and commerce abuses. Subsequent meetings of these companies resulted in a formulation titled "Ethical Principles for Offering Internet Health Services to Consumers." Members agree to follow the fourteen guidelines concerning the information presented on the site and the use of information gathered by it. (For a complete listing of the guidelines, go to (www.hiethics.org/Principles/index.asp)).

TRUSTe. This independent, nonprofit group attempts to build consumer trust and confidence in the Internet by promoting disclosure and informed consent. Approved sites display the TRUSTe icon, which signifies that the approved Website has agreed to notify users of Personally identifiable information collected from the user through the Website

The organisation collecting the information
• How the information is used
• With whom the information may be shared
• What choices are available to the user regarding collection, use, and
• distribution of the information

The security procedures that are in place to protect against the loss, misuse, or alteration of the information under the Website's control.

For more information about TRUSTe, see its Website: (www.truste. org/).

Health on the Net (HON). The health on the Net Foundation Code of Conduct (HONcode) for medical and health Websites addresses the reliability and credibility of information on the Internet. As posted on HONs Website, "The HONcode is not an award system, nor does it intend to rate the quality of the information provided by a Website. It only defines a set of rules to: hold Website developers to basic ethical standards in the presentation of information; and help make sure readers always know the source and the purpose of the data they are reading." For further information, see (www.hon.ch/HONcode).

Codes of ethics

Professional codes have evolved throughout history as practitioners grappled with various ethical and bioethical issues. Increasingly, groups of professionals have defined how members of their profession ought to behave. The Geneva Convention Code of Medical Ethics, established by the World Medical Association in 1949, is similar to the Hippocratic oath. This code refers to colleagues as brothers and states that religion, race, and other such factors are not a consideration for care of the total person. The Declaration of Helsinki, written between 1964 and 1975, is an update on human experimentation. it includes guidelines for both therapeutic and scientific clinical research.

Malpractice and risk management

Because e-health, telehealth, and telemedicine add a new element to care delivery, many providers question the level of liability when using technology. In medical practice, technology can thus add new and unforeseen factors to health care. For example, email offers new ways of interacting with patients and unique challenges that create new liabilities. Nevertheless, it is important to draw this distinction: telecommunication technology in health care offers a variety of new *tools* for extending existing specialities, but it does not represent a new speciality itself[4]. Therefore, discussing ethics, standards, malpractice, and technology risk management should avoid trying to reinvent each discipline. The goal is to examine how existing codes need to accommodate the inclusion of each telehealth tool (telephone, email, chat room, fax, videoconferencing, and the like) as a specific means of communication for specialised clinical care, education, or research.[6]

This chapter reviews some of the most pressing issues related to technology-based malpractice and risk management for both professionals and organisations. Although these areas are fraught with controversy, some groups have issued either preliminary or substantive papers to help make sense of the confusing landscape. We have attempted to synthesise the information from these sources in the form of suggestions to facilitate your decisions concerning the implementation of e-health, telehealth, and telemedicine.

Malpractice

Medical malpractice is generally defined as "a deviation from the accepted medical standard of care that results in injury to a patient for whom a clinician has a duty of care".[7]

For practitioners using rehabilitation technology, this definition presents complicated legal issues for both intrastate and interstate practice. Law generally governs malpractice liability, which defines the duty of care, the amount of damages patients may collect, and statutes of limitations.

Telehealth practise complicates issues of malpractice. One of the most challenging areas to define is the altered practitioner–patient relationships in case they are mediated through technology (Information communication and technology). This type of practice increases the risk when examining patients remotely without face-to-face contact between patient and practitioner. This risk can be reduced by having another professional present with the patient during remote consultations using technology. In many settings, a nonspecialist (the referring professional) makes a video call to the specialist (the consulting professional) and remains in the room with the patient. The nonspecialist performs the patient's physical examination at the specialist's direction and discusses findings throughout the procedure. This technique allows the specialist to make a diagnosis and guide intervention with immediate and systematic information from the nonspecialist. A nurse practitioner, for example, can efficiently perform the functions needed for most general exams required by a wide range of physicians. Despite the apparent effectiveness of such an arrangement, this practice could be regarded as delivering less than adequate care in some cases. Even with another professional present in the room, if something goes wrong, the burden of proving the effectiveness of the remote treatment falls upon the directing specialist. In addition to this problem, many practitioners do not typically operate with another professional in the room. One can see that video connections may be helpful for some patient populations, but it is also conceivable that they will not benefit others. Empirical research on the impact of various technologies on various patient populations is clearly needed to determine the efficacy of different technologies and the different liabilities associated with these technologies. For example, problems might arise with the use of compressed video, in which repetitious information is eliminated as the data are converted from analog to digital and back. The default setting on such technologies thus may be viewed as interfering with the diagnostic procedure, and the practitioner may be seen as rendering a diagnosis with incomplete information.

Despite these possibilities, telehealth may decrease the threat of malpractice suits by allowing for databased informational resources and better record keeping—such as videotaping consultations.

There are many advantages and disadvantages to video recording a consultation. A videotape automatically provides proof of the encounter,[8] however, such recordings may not be helpful if hindsight raises questions about decisions made by the reasonably cautious practitioner. Other unforeseen complications may arise. For example, creating audio, video, or other image records requires the patient's written consent.

The decision to make such recordings necessitates the additional steps of developing and using a thorough consent agreement. Elements to consider in such an agreement are detailed at the end of this chapter. Furthermore, although security measures are likely to be undertaken to protect the confidentiality, recordings may nevertheless be vulnerable to tampering or mishandling. Practitioners and organisations should consider the liability risks of maintaining and storing such records.[9] Research and case law will undoubtedly.

References

1. World Health Organization. A Health Telematics Policy in Support of WHO's Health-For-All Strategy for Global Health Development: Report of the WHO Group Consultation on Health Telematics. 1998. Available online: http://apps.who.int/iris/bitstream/10665/63857/1/WHO_DGO_98.1.pdf[Reflist] (accessed on 15 September 2021).
2. Curfman A, McSwain SD, Chuo J, et al. Pediatric Telehealth in the COVID-19 Pandemic Era and Beyond. *Pediatrics*. 2021;148 5.
3. Keenan AJ, Tsourtos G, Tieman J. The Value of Applying Ethical Principles in Telehealth Practices: systematic Review. *J Med Internet Res*. 2021;23:e25698.
4. Stamm BH, Pearce FW. Creating virtual community: telemedicine and self-care. In: Stamm BH, ed. *Secondary Traumatic stress: Self-care issues For clinicians, researchers, and Educators*. Lutherville, MD: Sidran Press; 1995:179–207.
5. Tavani H. *Ethics and Technology: Ethical Issues in an Age of Information and Communication Technology*. 2nd ed. USA: John Wiley & Sons; 2007.
6. Maheu M, Callan J, Nagy T. (in press). Call to action: Ethical and legal issues for behavioral telehealth including online psychological services. In: Bucky S (Ed.), Comprehensive textbook of ethics and law on the practice of psychology. New York: Plenum Publishers.
7. DeVille KA. *Medical Malpractice in Nineteenth-Century America: Origins and Legacy*. New York, NY: New York University Press; 1990.
8. Care Primary. *America's Health in a New Era*. Primary Care: America's Health in a New Era. Washington, DC: National Academy Press; 1996.
9. Marlene M, Allen A, Whitten P. E-Health, Telehealth, and Telemedicine: a guide to startup and success. John Wiley & Sons, 2002.

CHAPTER 3

Micro electrical mechanical system (MEMS) sensor technologies

Introduction to micro electrical mechanical system (MEMS)

Micro-electromechanical systems (MEMS) employs process technology to create tiny integrated devices or systems which combine mechanical and electrical components. They can be fabricated using integrated circuits (IC) batch processing techniques and range from few micrometers to millimeters. They have ability to sense, control and actuate on micro scales that generate effects on a macro scale.[1] The device electronics are fabricated using 'computer chip' IC technology, with the micromechanical components fabricated by sophisticated changes in the silicon and its substrates using micromachining processes. With the help of bulk, surface micromachining and high-aspect-ratio micromachining (HARM) selective parts of the silicon are either removed or added to the structural layers to form the mechanical and electromechanical components. While integrated circuits are designed to exploit the electrical properties of silicon, MEMS takes advantage of either silicon's mechanical properties or both its electrical and mechanical properties.[2] Usually the MEMS consist of mechanical microstructures, microsensors, microactuators and microelectronics, all integrated onto the same silicon chip.

Micro Electrical Mechanical Systems or MEMS were derived from semiconductor electronics and integrated circuit technologies. In the 1950s, optical lithography was used in printed circuit fabrication and semiconductor etching for the production of transistors. MEMS theories were formulated in the 1950s by experts from Bell Laboratories. They developed piezoelectric sensors employing silicon and germanium material whatever were used in the growth of sensor and actuator technology. In the 1980s precision technologies for mass production of microelectronics and integrated circuits were employed. From 1980 to the 1990s, the MEMS technology developed at a rapid pace using micromachining techniques to fabricate mechanical, electrical, chemical, biological, and medical devices having a small size,

Modern Intervention Tools for Rehabilitation.
DOI: https://doi.org/10.1016/B978-0-323-99124-7.00003-1

higher performance, and lower unit costs. This led to widespread interest in several fields by observing the immense potential about the technology.[1,2]

Large resources were invested by several countries post-1990s for growth and study of MEMS technology. Japan invested a huge sum of money in MEMS technology in the 1990s for the monitoring and repair of underground pipelines, blood vessels, and respiration channels. MEMS progressed from laboratory research to the development of industrial products. In the 21st century, MEMS is integrated with nanotechnology and smart materials to develop better high-end technological devices and gadgets. The technology was extensively employed to satisfy the needs of society, improve societal well-being, and raise the standards of living among people in the community. MEMS technology found applications in several fields such as automobiles, aeronautics, information technology, biomedical engineering, process control, and automation, telecommunication, environmental designs, consumer/home electronics, science, and technology, etc. which led to its quick widening with favour in almost all known fields.[3,4]

MEMS is an engineering technology whose scientific base is different from biology and medicine. To develop MEMS-based medical devices collaborative teams from different fields are required to share expertise and skills in their development. The MEMS technology was widely used in the development of smart healthcare devices and rehabilitative devices which has been explained in detail in the following section. The future holds great promise as the technology would expand in all fields with newer applications opening up with advancements in scientific know-how and technical insights soon. It could be employed to fulfill several unmet human needs such as automation in the field of agriculture, protection, and regeneration of natural resources, personalized healthcare, safety, automated home or consumer electronics, etc. The ease with which the MEMS products can be incorporated into microelectronics, integrated circuits, and computational devices for developing small volume, low power consumption, online devices, and quick response equipment(s) add to its growing popularity and usage.[5]

MEMS encompasses interdisciplinary fields such as manufacturing, design and engineering which includes circuit fabrication technologies, material science, electrical and chemical engineering, mechanical engineering, fiber optics, instrumentation etc. Due to complex design of MEMS they have wide range of applications across different products and markets. MEMS are employed in automotive designing, healthcare devices, electronic communication and large scale defence applications. They are used in blood pressure sensors, computer spare parts, biosensors, optical switches and many other products having high market value and potential.

In future, MEMS would be one of the most promising technologies for the 21st century with huge potential to revolutionize both industrial and consumer products by combining silicon based microelectronics with micromachining technology. It would affect the human lives and allow them advanced technologies to meet their daily needs. After semiconductor revolution, MEMS is potentially seen as second revolution.[6,7] The ultimate goal of MEMS technology is to satisfy user's needs and improve their living conditions. Its applications can be employed in all fields of human activities and daily requirements such as automotive, aeronautical, military, information technology, healthcare, telecommunication, environmental, consumer etc. The decision to utilize the MEMS technologies in the near future would depend on the ability to provide better performance, fulfill unmet market needs, efficiency, reliability of device and in specific industrial usage. The use would also depend on overall integration, packaging, interface circuits and computing or signal processing circuits. MEMS has advantages as compared with other traditional technologies. MEMS products, when incorporated with microelectronics, integrated circuits, and as computational systems, can be made into compact volumes, requiring less power usage, on–line analysis, and quick–response equipments to satisfy the needs of biological research and clinical care.[8,9]

Applications of MEMS in engineering product designing

Application of MEMS in rehabilitation

Since past few decades public interest and awareness concerning physical health and well-being with concern about surrounding environment activities has created an emerging need for smart sensor technologies and monitoring devices which can record, classify and provide feedback to the users. Such information is considered to be valuable and reliable for the overall health of the individual. Such devices have proved to be quite beneficial for patients with chronic illness requiring round the clock monitoring. Micro Electrical Mechanical Systems or MEMS devices play a vital cog in developing and designing such smart healthcare sensing devices which have huge market potential and commercialization opportunities in the near future.[5,6] A significant indicator of the massive trend in MEMS sensor technologies in the healthcare and well-being domain is represented by the continuous emergence of novel medical devices. The market of disposable medical devices embedding MEMS sensors for monitoring and diagnosis is forecast to rise more than 6 billion dollars in 2018. The use of sensors to monitor chronic diseases, such as hypertension, obesity, diabetes, sleep disorders and heart failures, is the key-element to maintain high the quality of life (often predicting the event) and also to reduce the cost of healthcare thanks to a remote monitoring; moreover, early intervention is vital for patients at risk of developing chronic diseases.[7,8]

Besides the extensive usage of the MEMS technology in the aeronautical field, it can serve as an important tool in biomedical and clinical research where light-weight, fast-response, miniature size, and ease to use the equipment are required which cannot be met with the traditional technologies. They can be easily incorporated into semiconductor devices, integrated circuits, microelectronics, and other computational devices. Since it is persistent to diagnose and cure the diseased human body, which would generate huge curiosity and interest in the growth of the technology that would have a high impact with several potential applications. Since the growth of MEMS technology in the healthcare field requires efforts from skilled collaborative teams having expertise in multi-disciplinary fields, the growth of this technology in the medical sector lagged in other fields, however, now with the availability of the desired resources and skilled manpower the field is ever-expanding in the healthcare sector. The growth of Bio-chip and DNA therapy allowed the MEMS technology to further progress in the healthcare field with the availability of skilled manpower. The biomedical applications of MEMS have been in three fields largely[1,2,6]:

a. Research in biology, medicine, and pharmacology to understand the DNA sequence and find the relationship of genome sequence which require the development of new technology and research facilities.

b. Individualized or personalized medicine with the aid of information technology and computing devices to adjust the treatment as per individual needs and available conditions.

c. Improved medical care to patients to measure physiological parameters and provide diagnostic care. They are being extensively employed in microsurgery, minimally invasive procedures, implant prostheses, tissue engineering, and electrical stimulation therapies.

The MEMS technology has been used for the evaluation of the rehabilitative process, detection of diseases, or accessing the impact of training activities. The technology can help stroke patients who suffer functional changes in the sensory, neurological, and musculoskeletal systems. To study the loss of motor function in upper and lower body extremities. In rehabilitation, the evaluation of Range of Motion (ROM) parameters is important means for determining clinical practice by employing several measurement techniques such as goniometer, inclinometer, and visual estimation of passive ROM or still photographs. An increase in the use of microelectromechanical systems (MEMS) in all sorts of consumer electronics, wearable devices, and sensors without affecting the daily routine functioning of an individual can be a boon for patients who are dependent on such devices for their daily routine activities.[7,9]

The obvious advantages of MEMS sensors are that they are low–cost, reliable, small, light, non–obtrusive, and can perform long-term measurements. MEMS sensors play a key element in innovative solutions in the vast majority of fields, including the measurement of human biomechanical parameters, such as ROM. They have the appropriate methodology and better outcomes for sincere users. It is imperative to deploy controlling tools that are required during setting task performance to obtain representative values otherwise inappropriate values could lead to incorrect diagnosis or mistaken evaluation of the patient's rehabilitation progress. The dimensions of MEMS devices vary from one micrometer to a few millimeters. They are employed in the development of Accelerometers, gyroscopes, magnetometers, pressure sensors, etc. which extensively employ MEMS technology.[3,5,7]

Recently, they are widely used in smartphones, 3D controls, pedometers, wearable sensors, inertial navigation, and even in targetable ammunition. MEMS sensors have their analog to digital (A/D) converter and are used in motion analysis using wearable devices. MEMS sensors are aimed at developing intelligent household environments and monitoring specific groups such as elderly, disabled, and physically challenged persons to monitor their physiological parameters such as blood pressure, motion and blood sugar, etc. Further, they can be employed in preventing any unwanted incidents such as fall prevention shoulder tilt and false gait, etc. The system control in the wearable devices should work efficiently and quickly to avoid harm to the patient if dangerous states are sensed. Wearable sensors are subject to research and development by interdisciplinary teams to improve the quality of life of people with disabilities and decrease their dependence on care providers.

The MEMS is employed for the study of neurons from a single cell unit to the scale of large populations and would continue to be an important tool for the neuroscience community. Mapping brain activity is expected to improve disease treatment and allow for the development of important neuromorphic computational methods. An important aspect of brain mapping technology is that it should be compatible with behavioral studies and require considerable advances in miniaturization and packaging. Neuromorphic computing, pattern recognition with high processing and computing power to achieve brain-like energy efficiency and adaptability. The need to avoid adverse tissue response, strain due to material mismatch, long term reliability. MEMS and microsystem engineers work closely with neuroscientists to develop devices employing reverse engineering for brain memory or learning.[6,8]

In the rehabilitation field, MEMS technology is employed in functional electrical stimulation (FES), artificial limbs, hearing aids and cochlear implants, heart pacing devices, epileptic seizures, hand gripping, mobility stimulation, suppression of pain, visual and neural prosthesis, etc. With the help of MEMS sensors different human biomechanical parameters could be studied which provide the range of motion that can serve evaluation for rehabilitation process, diseases detection and assessing the training needs of individuals. MEMS sensors are one of the most important innovative tools that are employed in study of human biomechanics and having potential tele-rehabilitation applications.[7] The enhanced usage of MEMS sensors in all consumer electronics led to the development of new kinds of devices and sensors with good sensing performance and development of smart wearable healthcare monitoring. This led to the development of smart, low-cost, reliable, non-obtrusive devices for long term monitoring with improved range of motion (RoM) sensors. The MEMS technology is slowly becoming into a significant clinical innovative tool with development of hardware and software platforms for at home rehabilitation and continuous patient monitoring.[8,9] This allows better physician and patient interaction with enhanced diagnosis and monitoring.

With increased public awareness concerning healthcare, physical activities, safety and environmental sensing has created an emerging need for smart sensor technologies and continuous monitoring devices able to sense, classify and provide user feedback about their overall health and personal well-being in a pervasive, accurate and reliable fashion. Such monitoring and accurately quantifying the physical activities of the users has proven an important aspect in the health management of patients affected with chronic diseases such as stroke, Parkinson's disease, Alzheimer, Autism etc. It provides an easy way for healthcare professionals as well the kin of these patients to monitor their regular health condition and provide requisite support when required. The MEMS technology in the field of Robotics would help in different ways such as providing sensors and actuators, development of smart intelligent autonomous systems and producing micro-robotics devices. This would help in development of smart rehabilitation based devices which would support better patient care and quality treatment opportunities. Lot of research and work is going on in the field and slowly new technological advancements are manifesting the field of robotic based rehabilitation. This would allow providing quality care to the elderly, specially abled and the vulnerable groups in the society making them independent and improving the societal status.

MEMS at microscopic scale

The MEMS technology allows transforming traditional large mechanical systems into miniature, better performing, and with high mass-producible alternatives, analogous to what the microcircuit and semiconductor technologies have done to the electrical and electronics system. Sensors such as MEMS accelerometers, MEMS gyroscopes, MEMS pressure sensors, MEMS tilt sensors, and other sorts of MEMS resonant sensors. Actuators like MEMS switches, micro-pumps, micro-levels, and micro-grippers. Generators and energy sources like MEMS vibration energy harvesters, MEMS fuel cells, and MEMS radioisotope power generators. Biochemical and biomedical systems like MEMS biosensors, lab-on-chips, and MEMS air microfluidic and particulate sensors. MEMS oscillators for accurate timekeeping and frequency control applications. At a good smaller Nano metric scale, the fabrication technology morphs into a nanoelectromechanical system (NEMS).[10] The MEMS are made up of components between 1 and 100 micrometers in size and their range ranges from 20 μm to a millimeter although when arranged in an array the size may range to more than 1000 mm^2.

Furthermore, where MEMS is integrated with other technologies, various combinatory embodiments can form, like BioMEMS where biochemical and biomedical systems are realized on micro-fabricated devices, micro-opto-electro-mechanical system (MOEMS), or Opto-MEMS where optical systems like micro-mirrors are integrated to manipulate or sense light at the microscopic scale, frequency microelectromechanical system (RFMEMS) typically involves close integration with semiconductor microelectronics to supply RF transduction and switching capabilities.[11]

The main disadvantage of employing MEMS is the fabrication and assembly unit costs could be quite high and expensive to develop for low quantity products. Therefore, MEMS are not suitable for niche applications unless cost is not an issue. Further, the testing equipments which may be employed to characterize the quality and performance of the developed MEMS device may be expensive and not feasible for use. The challenge of MEMS devices is rapid development and availability of low-cost thin wafers and the availability of the improved electrode material which shows exceptional electrical efficiency and processing in semiconductors.[12] MEMS is a manufacturing technology; a paradigm for designing and creating complex mechanical devices and systems as well as their integrated electronics using batch fabrication techniques.

Technological advancement of MEMS

The first laboratory demonstration of MEMS devices came about in the 1960s in the form of a MEMS pressure sensor. Academic research gained momentum within the 1980s, while commercial development and manufacturing took off within the 1990s. Today, everyone carries MEMS devices on themselves in the form of smartphones, smartwatches, and fitness trackers. Within the past, an aeronautic gyroscopic system won't to determine roll, pitch, and yaw within the cockpit of aircraft weighed several kilograms and measured several inches in length, whereas nowadays, MEMS gyroscopes in our smartphones weigh less than a milligram and are equivalent in size to a grain of sand. With miniaturization in size, also comes a big reduction in manufacturing cost and improved scales of economy. This is often like the continued miniaturization and reduction in cost seen within the semiconductor industry.[12] The Micro Electro-Mechanical System or MEMS is a chip based technology where sensors are composed of a suspended mass between a pair of capacitive plates. When sensor is displaced, difference in electrical potential is created by the suspended mass and measured by change in capacitance.

As long as quality IC design is frequently implemented as a series of discrete steps, MEMS design is much different; the design, layout materials, and packaging of MEMS are naturally associated. Whereas this, MEMS design can be more complicated than IC design – frequently required the same time development of every design 'Phase'. MEMS packaging is the process part of the perhaps most widely the CMOS design. MEMS packaging is meant firstly to protect the device from environmental harm while also providing an intersection and reducing unwanted external stress. MEMS sensors frequently use stress as a means of measurement; extreme stress can impair functionality by impairing the device and inducing sensor drift.[12,13] With future advancements in sensors technology and advance electronics the

MEMS based devices are going to become more sophisticated and concise to meet the different user requirements which would help in supporting the different activities which are presently difficult to achieve. There is no way the present technological advancements could hamper the growth of MEMS technology in our daily lives in the near future. These technologies would ultimately integrate with MEMS based sensors and provide a viable solution to the population and the community.

Application of MEMS in medicine and biomedical field

There is the various application of MEMS in Medicine and Biomedical. Technologies are growing in this field, and in the future, they will be a hugely important part of living beings. MEMS mechanics can be used to produce mechanical, electrical, fluid, thermal, optical, and magnetic structures also devices and systems. This technology is taken from the Integrated circuits Industry. MEMS in the biomedical field is creating a new generation of medical devices for the future.

Many fabrication techniques and materials are used to produce MEMS with the assist of the IC industry, also the development and purification of other microfabrication processes and non-traditional material took place. Photolithography, thermal oxidation, ion implantation, dopant diffusion, evaporation, wet etching, plasma etching, reactive-ion etching, ion milling also silicon dioxide, silicon, aluminum, and silicon nitride are conventional IC processes and materials. Additional materials are—micro-molding, batch micro-assembly, electroplating, X-Ray lithography, anisotropic wet etching of single-crystal silicon, thick film resist, spin casting, and PZT(Piezoelectric Film), Ni, Fe, Co (Magnetic Films); Sic and Ceramics (Temperature Materials), Stainless steel, Platinum, Gold, Sheet glass, PVC and PDMS (Plastics), etc.[14,15]

Two general methods for fabricating or integrating a complete MEMS device are there–
1. Surface Micromachining
2. Bulk Micromachining.

1. Surface Micromachining: External Micromachining is a method that produces MEMS by patterning, sedimenting, along with printing a series of narrow films (~1um thick).External Micromachining has been pre-owned to produce various MEMS devices for many applications & some of them are manufactured economically in large quantities.

2. Volume Micromachining: Volume Micromachining is different from surface Micromachining in the coating material whatever is single crystal

silicon that is designed and appearance to form the main part of the developing device. Using the expectable isotropic engraving attributing of single-crystal semiconductor, good deal of precision compound 3-D shapes namely V - channels, conic pits, membranes, vias, along with nozzles receptacle formed.[15]

The MEMS based technology would encompass several areas of healthcare technology which would improve the quality of care of being provided to patients presently that would allow their health conditions to improve and help in their early recovery. It would also allow their integration in the society and improve their overall well-being. The treatment procedures would become more affordable and within the budget of the lower income groups of the society. It would provide smart interface with healthcare providers for continuous monitoring of the health conditions of their patients. The technology would allow to provide immediate intervention to such patients when required or it is sensed by the MEMS based sensor. It would also reduce the length of their hospital stay, costs involved and make it more convenient for them to utilize the healthcare facilities in an optimum manner.

MEMS and drug delivery

Drug delivery systems play a crucial role in the treatment and management of medical conditions. Microelectromechanical systems (MEMS) technologies allow the development of advanced miniaturized devices for medical and biological applications. MEMS allows hundreds of hollow microneedles to be fabricated on a single patch of area 1 square of Centimeters. By applying this patch to the skin drug is delivered to the body by using micropumps. Micropumps are controlled electronically to allow a specific amount of the drug and also delivers at specific intervals. Due to the size microneedles are too small to reach and stimulate nerve endings, these don't cause any pain in the body. The majority of MEMS drug delivery systems consist of three components: drug chamber, drug release mechanism, and packaging. Drug is transferred from the drug chamber to a specific location in the body using a variety of actuation mechanisms that afford accuracy, precision, and reliability.

By the process of microfiltration, conventional filters are produced that are capable of screening micron-scale objects which results in an unacceptably broad statistical distribution of objects size that can pass. Micromachining and MEMS technology came to be worn for realizing filters that are uniformly and precisely machined that reducing the statistical variation in objects that passes through.[16]

The powered MEMS drug delivery devices can be classified as non-mechanical and mechanical. The non-mechanical micropumps transform non-mechanical energy into kinetic momentum and drive fluid out from the reservoir which can work at micro-sclae. The non-mechanical micro pumps do not require moving parts and hence structures are more simple and easy to fabricate. However, these actuation mechanisms are not suitable or have not been successfully used in drug delivery devices because their driving effect and performance (e.g., flow rate, response time, and pressure generation) are inferior when compared to mechanical actuation. Mechanical micropumps utilize moving parts to generate oscillatory or rotational pressure forces on the working fluid to displace it. Three movement mechanisms have been employed by mechanical micropumps: reciprocating, rotatory, and peristaltic. The majority of micropumps reported utilize reciprocating motion. This type of micropump requires a pumping chamber coupled to a physical actuator and a moving surface (diaphragm), and check valves to control fluid flow during the supply and pumping cycles. During operation, the actuator mechanism acts on the diaphragm resulting in an overpressure on the drug that displaces it from the pumping chamber. The majority of mechanical micropumps have fast response time, large actuation force, good biocompatibility, but are limited by high driving voltages and complex fabrication processes.[16,17]

Smart pill

A smart pill is a MEMS device that receptacle instrument in the human body. Smart Pill consists of Biosensors, Battery, Control Circuitry, and Drug reservoirs. The Biosensor made up from copper, magnesium and silicon senses the insulin levels within the body. The pill starts releasing the drug when the quantity falls below a certain amount required by the body. The signal is sent to an external device to detect the mixture of smart pill components within the body's digestive fluids. The smart pill can be ingested to record several diagnostic variables as it travels through the gastrointestinal (GI) tract to detect abdominal pain, nausea/vomiting, bloating, constipation or bacterial overgrowth. It can measure temperature, pressure in the muscles and pH levels while traveling through the GI tract. They could prove to be beneficial in treating cognitive dysfunction in patients with Alzheimer's disease, ADHD or schizophrenia patients. The smart pill costs close to $1500 and it can be easily excreted by the body within 24–48 h post usage.[14,18] Further research is ongoing in the development of the smart pills to make it more compact, affordable, and sustainable to allow targeted drug delivery to the patients. The future of the smart pill market is quite immense and profitable area for the industry to grow and prosper for meeting the unmet needs of the society at large.

Blood pressure sensor

Pressure sensors have been widely used in MEMS technology. To use a pressure sensor the person has to select the needed or right sensor from the various application. Pressure sensor manufacturers of MEMS have medical applications only for larger markets like blood pressure monitoring. The sorts of Parameters of the device include pressure range, pressure sensitivity, device impedance, device equivalent circuit, sensor size, pad size, pad placement, and other parameters.

Integrated Sensing System Inc. ISSYS performed studies on the BioMEMS of the wireless, battery-less, implantable pressure sensor. ISSYS is developing intelligent BioMEMS sensors and systems to occur the quality of medical treatment that can apply to measure pressure gradient, identification and detector clogged heart failure, measure cardiac output, and improve gastrointestinal tract diagnostic capabilities to treat the gastroesophageal reflux disease (GERD), diagnose the urological disorders, measure drug delivery rate for infusion systems.

The Cleveland Clinic Foundation BioMEMS laboratory is trying and growing the systems in Biomedical application since BioMEMS will bring smaller, more accurate, less invasive, and most cost-effective Biomedical devices. A mini implantable BioMEMS pressure sensor monitors monitor physiological measurements, such as abdominal, aortic pressure, and intraocular, intracranial, and intervertebral pressures. The majority of MEMS used in biomedical uses act as sensors, i.e. include analytical sensors used during surgery (i.e. measured intravascular blood pressure), long-term sensors for prosthetic devices sturdy experienced sensors and their uses are yonder the extent of here paper.

A lab-on- a chip (LOC) takes the laboratory testing of biomolecular samples like blood, urine, sweat, and sputum out of the clinical lab and places them in the field or point of care. LOCs use microfluidics and the chemical sensor's however to identify multiple analytes substances being analyzed.[16,17] With increase of heart diseases, risk of chronic diseases, sudden death and heart failure the continuous monitoring of blood pressure would allow the technology to further grow and develop in the near future which would make it more sophisticated and user based with convenience to use the device for long-term purposes. Hence, the market potential for development of smart MEMS based sensor technology is quite immense and would become highly popular among the users in the near future.

Advantages of MEMS

There are various advantages of MEMS with the growing technologies. Some of them are given below:
- MEMS devices acquire very low power consumption through their processing speed is faster as well as the functioning of this device is greater.
- MEMS switches and actuators also can attain very high frequencies. MEMS devices are low cost due to scalable in manufacturing.
- The accuracy of MEMS technology is higher as well as they are more sensitive and more selective in nature.
- Frequency of MEMS switches and actuators getting higher.
- Having micro-size or small size like 1-micrometer feature MEMS is applicable for many applications.
- MEMS devices can be easily altered without changing the entire device
- MEMS can be readily integrated with microelectronics and integrated circuits
- MEMS have improved reproductivity

Disadvantages of MEMS

- The development of MEMS device is quite expensive and requires extensive testing.
- Prior knowledge required for implementing MEMS design.
- Involves high cost and setting up expenses.
- Research and development costs are too high to design MEMS devices
- Quality and performance testing can be costly
- Fabrication and assembly unit costs can be very high for low quantities. Therefore, MEMS are not suitable for only niche applications, unless cost is not an issue.
- Prior knowledge required to integrate MEMS devices with other electronic components

- The MEMS design is too complex.[17,18]
- Common mode of failure of MEMS devices may be due to fracture, creep, wear and tear, degradation, contamination, electrode discharge etc.

Future growth of MEMS

MEMS-based products produced growth so far has come from a combination of technology displacement, as exemplified by automotive pressure sensors and airbag accelerometers and new products, like miniaturized guidance systems for military applications and wireless tire pressure sensors. Much of the growth in the MEMS business is expected to come from products that are in the early stages of development or yet to be invented. A number of these devices include disposable chips for performing assays on blood and tissue samples, which are now performed in hospital laboratories, integrated optical switching and processing chips, and various RF communication and remote sensing products. The key to enabling the projected 25-fold growth in MEMS products is the development of appropriate technologies for integrating multiple devices with electronics on one chip. At the present, there are two approaches to integrating MEMS devices with electronics. Either the MEMS device is electronics on a single chip. At present, there are two approaches to integrating MEMS devices with electronics. Either the MEMS device is fabricated in polysilicon, as a part of the CMOS wafer fabrication sequence, or a discrete MEMS device is packaged with a separate ASIC chip. Neither of these approaches is entirely satisfactory, though, for building the high-value, system-on-chip products that are envisioned. It's this author's opinion that a combination of self-assembly techniques in conjunction with wafer stacking offers a viable path to realizing ubiquitous, complex MEMS systems.

The recent success of additive manufacturing processes (also called, 3D printing) within the manufacturing sector has led to a shift in the focus from simple prototyping to real production-grade technology. The improved capabilities of 3D printing processes to create intricate geometric shapes with high precision and determination have led to their increased use in the fabrication of microelectromechanical systems (MEMS). The 3D printing technology has offered tremendous flexibility to users for fabricating custom-built components. Over the past few decades, varied types of 3D printing techniques have evolved.

Technologies are developed. This text provides a comprehensive review of the recent developments and significant achievements within the most widely used 3D printing technologies for MEMS fabrication, their working methodology, advantages, limitations, and potential applications.

Furthermore, a number of the emerging hybrid 3D printing technologies are discussed, and therefore the current challenges associated with the 3D printing processes are addressed. Finally, future directions for process improvements in 3D printing techniques are presented. The growth of 3C-SiC on (001) silicon substrates utilizing vapor phase epitaxy is described. The growth mechanisms are discussed with the aid of structural and morphological characterizations performed by X-ray diffraction, transmission electron microscopy, and atomic force microscopy. Raman spectroscopy was used to study the residual stress. A large shift of Raman peaks concerning the expected values for the bulk is observed and explained by the relaxation of Raman selection rules due to lattice defects. The stress and stress gradients through the film thickness are observed and studied on micrometer-sized structures such as membranes and cantilevers. Local Raman peak fluctuations are observed on millimeter-sized membranes, while cantilevers show different degrees of curling depending on film thickness.

Experimental results of a modified micro-machined microelectromechanical systems (MEMS) mirror for considerable improvement of the diagonal laser scanning performance of endoscopic optical coherence tomography (EOCT) are there. Picture deformation due to securing of MEMS reflector in our preceding design was analyzed and found to be attributed to excessive internal stress of the transverse bimorph meshes. The modified MEMS mirror eliminates bimorph stress and the resultant buckling effect, which increases the wobbling-free angular optical actuation to greater than 37°, exceeding the transverse laser scanning requirements for EOCT and confocal endoscopy. The new optical coherence tomography (OCT) endoscope allows for two-dimensional cross-sectional imaging that covers an area of 4.2 mm × 2.8 mm (limited by scope size) and at roughly 5 frames/s instead of the previous area size of 2.9 mm × 2.8 mm and is highly suitable for noninvasive and high-resolution imaging diagnosis of epithelium lesions in vivo. EOCT images of normal rat bladders and rat bladder cancers are compared with the same cross-sections acquired with standard bench-top OCT. The results demonstrate the potential of EOCT for in vivo imaging diagnosis and precise guidance for removal biopsy of early bladder cancers.

Advances in microelectromechanical systems (MEMS) technology have guide to the development of a multitude of new devices heretofore impossible. However, applications of these devices are still hampered by challenges posed by their integration and packaging. The current trend in micro/nanosystems is to produce ever smaller, lighter, and more capable devices at a lower cost than ever before. In addition, the finished products have to operate at very low power and in very adverse conditions while ensuring durable and reliable performance. Some of the new devices are developed to function at high operational speeds, and others to make accurate measurements of operating conditions of specific processes. Regardless of their applications, the devices have to be packaged to facilitate their use. MEMS packaging, however, is application-specific and, usually, has to be developed on a case-by-case basis. To facilitate advances in MEMS, educational programs have been introduced addressing all aspects of their development. This paper addresses progress in MEMS by presenting pertinent features in the development of MEMS including, but not limited to, design, analysis, fabrication, characterization, packaging, and testing. This presentation is illustrated with selected examples.[12,14]

With advancement in MEMS technology, it would become a device of everyday usage in the hands of people such as smartphone for routinely monitoring their health and carrying out their daily activities of living (ADL) tasks in multimodal cloud sensing mode. This would allow integrated management of large volume of data which would represent a smart network with advanced computational capabilities. The device would continuously monitor user and environment conditions. The only constraint being the energy requirements of these devices which require smart power management and signal processing system. The MEMS technologies should become more user oriented and provide better performance with capability being less expensive or invasive in use. This would involve combination of multitude technologies such as big data, cloud computing, high performance applications which require immense technological advancements and scientific acumen to translate the laboratory research into practical usage for everyday performance of desired tasks. However, the growth of MEMS technologies would accelerate significantly driven by growing commercial demand and management of ambient conditions to distinguish between false positive and false negative alarms which may prove decisive in the final analysis of the results.[16,18]

The usage of MEMS are significantly growing in the development of non-contact physiological sensing devices through enhanced biocompatible material flexibility which prevent variations in sensor recordings. However, their usage is limited in the measurement of chemical and gaseous sensing applications. However, with advancements in molecular based sensing technologies, capability for early detection of biochemical disease markers, improvement in semi-conductor technologies the usage of MEMS would significantly increase and they would become an efficient, reliable and stable future option in several applications including healthcare devices.[19,20]

Discussions and summary

Although the vend for MEMS devices is still in infancy, however, for it to see a boom like the Integrated Circuits (IC) market of the 1970s would require more sustained demand. As there is no dominant technology such as metal oxide semiconductor circuitry which accelerated the growth of the digital electronics industry. Although MEMS is an enabling technology as it helps in the development and production of several industrial and consumer products, it is considered a disruptive technology that requires a completely different set of capabilities and competencies to implement. Hence, critical technological bottlenecks, economic viability, and market scenario need to be understood before the MEMS technology could be employed in wider use. For the interfacing of MEMS devices, cost reduction is important with the availability of better infrastructure resources and more reliable manufacturing processes with requisite technical information of new standards on interfacing.[18,20]

For the development and growth of the MEMS technology, it is important that small total package volume, body-powered source, wireless communication, material biocompatibility to protect the implanted devices, and sensors that can work for a longer time duration without getting damaged by the body environment and can be fabricated with reasonable costs are considered for different industrial applications. Hence, concentrated effort and attention are required in the biomedical field for the growth of MEMS technology in near future.[19,21] The MEMS technology is still in its infancy and further research is required to make it commercially viable and application based. No doubt the technology is slowly manifesting in our daily life, however, the above stated factors needs to be addressed before it could be put to use in achieving long term sustainable goals in our lives. The future of the technology is quite bright and it would revolutionize the way present tasks are achieved without its involvement.[22,23]

Moreover, new approaches and techniques based on MEMS technologies have enhanced tremendously over the past 30 years with widespread applications in the field of wireless communications, sensor networks, rehabilitation, environment sensing, medical devices, electronic circuitry etc. The deployment of MEMS sensors in field of healthcare could significantly alter present healthcare paradigms with improved efficiency and reliability of such devices in better patient care and quality treatment in healthcare set-ups.[24,25] They can predict early healthcare signs and better manage those specific events before the problem becomes of alarming proportion. The MEMS technologies can make the healthcare more of "wellness preventive model" from present "reactive healthcare model". With cost of sensor technologies declining and better features such as storage, communication, data transfer etc. available in the near future, the role of MEMS technologies would significantly grow in all sectors of its application and more so in the healthcare sector.[26,27] Ultimately, we would see that all technologies would encompass the MEMs based technology or may be directly or indirectly linked with it, so that better outcomes could be possible with its usage in the near future. The market potential and dependency over such devices would enhance in the future if they become more technologically sound, viable, smart and sophisticated to respond to any requirements of the user, industry or the society. Hence, it may be concluded that market potential of these devices if quite enormous if they could be developed keeping in mind the needs of the society and their overall applications.

References

1. Karchňák J, Šimšík D, Jobbágy B, Galajdová A, Onofrejová D. MEMS sensors in evaluation of human biomechanical parameters. *Procedia Eng.* 2014;96(2014):209–214.
2. Patel S, Park H, Bonato P, Chan L, Rodgers M. A review of wearable sensors and systems with application in rehabilitation. *J Neuroeng Rehabil.* 2012;9(21):1–17.
3. Mukhopadhyay SC. Wearable Sensors for Human Activity Monitoring: a Review. *IEEE Sens J.* 2015;15(3):1321–1330.
4. Shany T, Redmond SJ, Narayanan MR, Lovel NH. Sensors-based wearable systems for monitoring of human movement and falls. *IEEE Sens J.* 2012;12(3):658–670.
5. Bora DJ, Kumar N, Dutta R. Implementation of wireless MEMS sensor network for detection of gait events. *IET Wirel Sens Syst.* 2019;9(1):48–52.
6. Ko WH. Trends and Frontiers of MEMS. *Sens. Actuator A Phys.* 2007;136(1):62–67.
7. Seymour JP, Wu F, Wise KeD, Yoon E. State-of-the-art MEMS and microsystem tools for brain research. *Microsyst Nanoeng.* 2017;3:1–16 16066.
8. Hochberg LR, Bacher D, Jarosiewicz B, et al. Reach and grasp by people with tetraplegia using a neurally controlled robotic arm. *Nature.* 2012;485(1):372–375.
9. Daponte P, et al. Design and validation of a motion-tracking system for ROM measurements in-home rehabilitation. *In Measurement.* 2014;55(2014):82–96.
10. Židek K, Dovica M, Líška O. Angle Measuring by MEMS Accelerometers. *J. Autom Mob Robot Intell Syst.* 2012;6(4):3–6.
11. Brand O. Microsensor integration into systems-on-chip. *Proc IEEE.* 2006;94:1160–1176.
12. Barbaro M, Caboni A, Cosseddu P, Mattana G, Bonfiglio A. Active devices based on organic semiconductors for wearable applications. Information Technology in Biomedicine. *IEEE Transactions.* 2010;14:758–766.

13. Sobek, D, Senturia, SD, and Gray, ML, Microfabricated Fused Silica Flow Chambers for Flow Cytometry, Proceedings of the Solid-State Sensor and Actuator Workshop, Hilton Head Island, SC, June 13–16, 1994, pp. 260–263.

14. Tjerkstra, RW, De Boer, MJ, Berenschot, JW, Gardeniers, H, Albert Van den Berg, Elwenspoek, MC. Etching Technology for Microchannels, Proceedings of the 10th Annual Workshop of Micro Electro Mechanical Systems (MEMS '97), Nagoya, Japan, Jan. 26–30, 1997, pp. 396–398.

15. Smith, L, and Hok, B, A Silicon Self-Aligned Non-Reverse Valve, Proceedings of Transducers '91, the 1991 International Conference on Solid-State Sensors and Actuators, San Francisco, CA, June 24–27, 1991, pp. 1049–1051.

16. Ahn, CH, and Allen, MG, Fluid Micropumps Based on Rotary Magnetic Actuators, Proceedings of the 8th Annual Workshop of Micro Electro Mechanical Systems (MEMS '95), Amsterdam, Netherlands, Jan. 29–Feb.2, 1995, pp. 408–412.

17. Liao M, Wan P, Wen J, Gong M, Wu X, Wang Y, et al. Wearable, healable, and adhesive epidermal sensors assembled from mussel-inspired conductive hybrid hydrogel framework. *Adv Funct Mater.* 2017;27(48).

18. Zhao X, Guo B, Wu H, Liang Y, Ma PX. Injectable antibacterial conductive nanocomposite cryogels with rapid shape recovery for non-compressible hemorrhage and wound healing. *Nat Commun.* 2018;9(1):1–17.

19. Ma C, et al. Balance Improvement Effects of Biofeedback Systems with State-of-the-Art Wearable Sensors: a Systematic Review. *Sensors.* 2016;16(4):434.

20. Masci I, Vannozzi G, Bergamini E, Pesce C, Getchell N, Cappozzo A. Assessing locomotor skills development in childhood using wearable inertial sensor devices: the running paradigm. *Gait Posture.* 2013;37:570–574.

21. Zanetti S, Pumpa KL, Wheeler KW, Pyne DB. Validity of the sensewear armband to assess energy expenditure during intermittent exercise and recovery in Rugby Union players. *J Strength Cond Res.* 2014;28:1090–1095.

22. Bassetti M, Braghin F, Castelli-Dezza F, Negrini S, Pennacchi P. Sensor nodes for the dynamic assessment of Alpine skis. In: *Topics in Modal Analysis II.* Berlin, Germany: Springer; 2012:471–479.

23. Depari A, de Dominicis C, Flammini A, Rinaldi S, Vezzoli A. Integration of Bluetooth handsfree sensors into a wireless body area network based on smartphone. In: *Sensors.* Berlin, Germany: Springer; 2014:547–551.

24. Domaneschi M, Limongelli M, Martinelli L. Multi-site damage localization in a suspension bridge via aftershock monitoring. *Ing Sism.* 2013;30:56–72.

25. Salarian A, Horak FB, Zampieri C, Carlson-Kuhta P, Nutt JG. Aminian K: iTUG, a sensitive and reliable measure of mobility. *IEEE Trans Neural Syst Rehabil Eng.* 2010;18:303–310.

26. Weiss A, Herman T, Plotnik M, et al. Can an accelerometer enhance the utility of the Timed Up & Go Test when evaluating patients with Parkinson's disease? *Med Eng Phys.* 2010;32:119–125 105.

27. Salarian A, Russmann H, Vingerhoets FJ, et al. Gait assessment in Parkinson's disease: toward an ambulatory system for long-term monitoring. *IEEE Trans Biomed Eng.* 2004;51:1434–1443.

CHAPTER 4

Brain stimulation in rehabilitation

Introduction

Brain stimulation is the process to stimulate both chemical and functional change in neuron level that helps to reactive brain functioning. Brain is the most complicated part in the human body. In human body brain is only organ who is responsible for thoughts processing, emotional coefficients, bodily movements, feeling, memories, sensations and all behaviors.[1] Human brain is the complex organ where various neurons and neurotransmitters works together. Brain cells constantly perceiving the information from the environment and react accordingly. Neurotransmitters work as a chemical messenger as in excitatory and inhibitory forms.[2] Which stimulates the electrical activities and calming the activities and helps the system to be work as balancing form. In the process of normal brain functioning the brain cell receives the information in form of electrical impulses that travels through the axon and axon terminal where neurotransmitters work as a chemical messenger. Brain stimulation whether its invasive or non-invasive helps to treat various mental and neurological disorders. Brain stimulation rehabilitation involve either activating or inhibiting the brain directly with electricity.[3]

Brain stimulation rehabilitation (BSR)[4] is a clinical care given by healthcare professionals that helps you get back, keep and improve abilities that you need for daily life. These abilities can be mental, physical or cognitive which is lost due to some illness or due to any injuries. BSR helps in improving quality of life for daily leaving.

Brain stimulation technologies

Brain stimulation helps in neural modulation for a specific brain region by electromagnetic field. Stimulation can be direct or indirect. Direct by mean of implantation of electrode and indirect means without impanation of electrodes.[5] Brain stimulation technologies can be invasive or non-invasive. With the help of these techniques brain activities stimulate new passages between nerve cells and help the brain to generate or develop new cells[6] which is called as neuroplasticity. It can be invasive and noninvasive. In

Modern Intervention Tools for Rehabilitation.
DOI: https://doi.org/10.1016/B978-0-323-99124-7.00004-3
45

invasive the surgical intervention is required. In this intervention electrodes are implanted into the skull, and it generates the electrical impulses that control abnormal brain activity. These electrical impulses we can adjust for the chemical balancing and helps to improve many neurological conditions.[7] In brain stimulation system the electrode or lead is insulated with a fine wire and implanted into a specific brain area where the stimulation is required to enhance the brain activities. The extended insulated wire of generator passed under the fulcrum of head and connecting the electrode to the internal pulse generator. In deep brain or in invasive brain stimulation many clinical conditions such as dystonia, epilepsy, Parkinson's disease and many brain disorders can be treated. This technique is very much effective in the treatment of movement disorders. This technique has some risk factor also such as risk of brain hemorrhage, infection, malfunctioning of device, headache, worsening mental and emotional status. In noninvasive process during TMS stimulation electromagnetic coil delivered the electromagnetic waves in pain free manner and helps in neuroplasticity.[8] TMS have many benefits such as it doesn't cause seizures, any sedative effect or any anesthetic effects after stimulation. Similarly, like other stimulation it also has some side effects such as headache, scalp discomfort at the site of stimulation, tingling and twitching of facial muscles. Some of the brain stimulation techniques are as following.

Vagus nerve stimulation

Vagus nerve stimulation involves use of electrical device to stimulate the vagus nerve with electrical impulses. Implanted vagus nerve stimulation is approved from food drug administration (FDA) to treat various neurological and psychiatric condition such as epilepsy and depression. This technique was originally developed for the treatment of epilepsy but latter on its was discovered that it also effects mood and other brain activities in positive manager for the treatment purpose for many psychiatric disorders.[9] In vagus nerve stimulation a device called as pulse generator is surgically implanted in the upper left side of the chest. This device is very small in size like stopwatch. The pulse generator has electrical lead wire that connect the generator to the left vagus nerve. In this technique vagus nerve carry massages from the brain to the body and it controls the mood, sleep and other functions.

Mechanism of action of vagus nerve stimulation

Vagus is the tenth cranial nerve which is arises from rootlet of medulla[10] it carries both efferent as well as afferent fibers.[11] It has multiple potential mechanism as well as efficacy to treat neurological conditions. In epileptic patients it results the cortical desynchronization[12] and control the abnormal hyperactivities in the brain by changing the activities of neurotransmitter such as GABA, serotonin and norepinephrine.[13] Which help in improving the functional activities of brain region and enhances the synaptic plasticity.[14] It also has some adverse effect which is depends on many two things where vagaus nerve stimulation is required surgical intervention[15] or its required stimulation only, so adverse effect varies on patient condition and on severity of disease.

Transcranial magnetic stimulation

Transcranial magnetic stimulation (TMS) is a non-invasive method in which externally placed, rapidly changing magnetic field causes induction of weak electric currents that lead to changes in neuronal polarization and activity. In another word we can say TMS is the non-invasive stimulation of the human brain with the help of magnetic pulse through a coil which need to be navigate on human cranium. These coil of TMS can be figure of eight or circular it depends on the brain area which need to be stimulate. TMS is approved by food and drug administration (FDA) USA, for many neurological and psychiatric conditions. In 2008 TMS was first approved for the treatment of depression. Faraday et al. in 1932 first discovered about electromagnetic induction and give the concept of brain stimulation. In 1985 Barket et al. modulated it into the electrical stimulation and develop the concept of brain stimulation. Barket was the first who designed it. The working mechanism of TMS is based on Faraday's low of electromagnetic induction. Faraday low of electromagnetic induction stated that magnetic field near the conductor transfers into the electrical current and do the depolarization of underlying nerve cells as shown in Fig. 4.1 (Evolution in TMS).

Mechanism of transcranial magnetic stimulation

TMS is worked on principle of electromagnetic induction in which magnetic field is produced that penetrates the cranium virtually and painlessly. When we applied TMS it results muscle contraction in the contralateral side.[16] In TMS small electromagnetics coil which is controlled by computer program, and it deliver short, powerful bursts of magnetic energy in the targeted area. The coil of TMS is applied over the cranium which target the focus area of treatment. TMS do, not affect whole brain it only reaches 2–3-centimeter depth of the cranium. The single pulse of TMS is used to

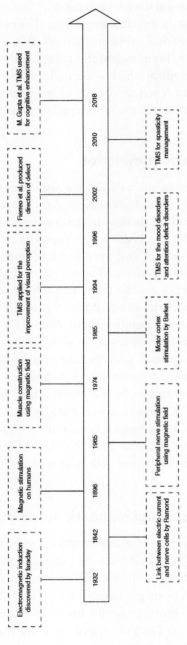

Figure 4.1 Evolution in TMS.

calculate motor evoke potential (MEP).[17] MEP is elicited from the motor cortex when it passes through the scalp in the targeted muscle. As magnetic field moves into the brain it produce small electrical current.

Transcranial electrical stimulation (tES)

Transcranial electrical stimulation(tES) is the recent advancement of device which is used for the neuroscientific research including transcranial direct current (tDSC) and alternating current (tACS). tES is also noninvasive brain stimulation technique in which micro electrical current passes through the cortex of the brain and stimulated the brain functioning. Which helps to alter the brain neural activities by excitation and inhibition of neuronal charges. tEC are basically two type it can be transcranial direct current (tDSC) and alternating current (tACS). In tDSC application we generally using two conductive rubber electrodes which is soaked in saline water, and which connected to the head with nonconductive elastic straps. Another electrode is placed on cortical area to modulate the neural activities. During the tDCS the low intensities current propagate through the head and returned via the reference electrode and this process modulate neuronal transmembrane potentials and modulating the excitability of underlying neurons. In tDCS the intensity of around 1 mA induces the long-lasting changes with homogenous DC field. Which helps to manipulate the brain excitability via membrane potentials. Various other parameters also influence the effect of tDCS. In tACS generally sinusoidal wave are used for the stimulation purpose. it does not alter neuronal excitability but entrains the exogenous frequency. In tADC neuronal entrainment is achieved by applied current that polarized the neurons reflections.

Safety and tolerability of tES

In last past decade various noninvasive brain stimulation techniques have emerged as visible, safe and low-cost method for the treatment of various illness. tES is a neuro-modulatory technique which involves application of weak direct current with the help of electrode over the scalp region where the stimulation is required. The device does not have any serious side-effect but some moderate adverse effect such as skin burning, skin burning, skin irritation headache and fatigueless are the common side effects.

Evolution in TMS

Repetitive transcranial magnetic stimulation (rTMS) is an as of late evolved painless non-invasive brain stimulation technique for the treatment of neurological and psychiatric problems. It includes the use of changing magnetic field to the superficial layers of the cerebral cortex, which contains little electric currents, called as 'eddy current'. In this circumstances, cerebral cortex work as a secondary coil.[18] TMS enjoys a benefit as it is concentrated and bypass the impedance of skull and superficial layers. Even though, its mechanism of action is still not revealed, current proof focuses on its part in causing long term excitation and inhibition of neurons in specific cerebrum regions. Evidence in support of rTMS as therapeutic tool continues to grow, there is a need to develop standardized protocols for its organization.[19] It does not cause any major reactions with rTMS, however its utilization is confined in those having magnetic implants or neurological or heart problems. Among all the indication of rTMS, the proof is more towards the treatment of refractory unipolar depression.

The development of modern transcranial magnetic stimulation

Observations on electrical excitation of cerebral cortex began in 1874 in which contralateral motor feedback was evoked. In 1881, Faraday gave the principal on electro-magnetic induction.[20] But Kolin et al.[21] in 1959 accomplished nerve excitation by involving magnetic energy in frogs which established the groundwork for EM stimulation of brain tissue for diagnostic and therapeutic purposes. In last couple of years, rapid progress has been made in the development of coil shapes to concentrated magnetic field to attain better command over the spatial extent of excitation. While the old type of treatment took up to 37 minutes per sittings, with high-frequency (HF) theta-burst stimulation the sittings may only last a few minutes.[22] It is logical that treatment conventions will go through additional refinements in the coming years to make it more comfortable for patients.

Neurostar

It is the first rTMS Machine was approved in 2008. It required 13 years and many clinical trials for TMS to be supported by FDA in the United States. Neuronetics Inc made the first Transcranial Magnetic Stimulation, and it is known as NeuroStar. NeuroStar is an extraordinary, featured machine, and as a rule, they have the most acceptance among patients and experts, because of plugs like these where people are told to "ask for NeuroStar by name". Among all the TMS machines, neurostar is the first surface, implying that they excite the external cortex of the brain. It is also known as "rTMS". rTMS just means more than one pulse for per sittings, to treat depression. Thus, NeuroStar act as both the rTMS machine and Surface TMS machine.

Deep TMS

The second machine which was approved by FDA was made by an Brains way. They guaranteed that their machines than NeuroStar. Brains way stimulator utilize a huge helmet that is put on a patient's head during the entire sitting. Deep TMS has not been demonstrated that the Brains way machines really hit a deeper part of the brain. Notwithstanding, Brains way's own round models assumed that their Deep TMS gadgets arrive at structures 0.7 cm deeper than different machines. One of the large benefits at first for utilizing Brains way machines was that the treatment length was more limited, controlled in simply 20 min contrasted with 37.5 min for different TMS machines. This new protocol permitted doctors to diminish the inter-stimulation interval by about half so the treatment should be possible in close to a half of the time.

Neuronavigated TMS (FMRI TMS)

A few TMS hospitals might send to get a MRI before starting treatment. For this situation, the expectation is that the MRI will tell precisely where to put the stimulator on the scalp to arrive at the left dorsolateral prefrontal cortex. This strategy has to some degree become undesirable as it doesn't appear to have any significant effect in viability contrasted with the ongoing consensus algorithm for finding the spot. The prefrontal cortex is really one of the biggest parts of the brain, so there is a lot more space for errors than initially recollected in 2008. The actual MRI can be very expensive, uncomfortable and time consuming.

Theta burst stimulation

Theta Burst Stimulation, also known as Express TMS or TBS. In TBS, the stimulator pulses at a particular frequency that copies ones' own brainwaves, empowering the cerebrum to enhance the neuroplasticity up to multiple times quicker than conventional TMS. While a TMS convention used to require 37.5 min, theta burst stimulation[23] comes by precisely the same outcomes in only 3 min. Theta burst was FDA approved in 2018 after it was shown that both the viability and harmful effect profiles were equivalent to standard 10 Hz[24] rTMS treatment. Theta burst is preferred as it is by all accounts a general sign for the mind to draw in and increases brain neuro-plasticity. Old TMS conventions have a ton of individual change abilities: it is an excitatory stimulus for patients, in that it will enhance the action, however it can likewise be an inhibitory stimulus for others, implying that a specific subset of individuals can really seek more regrettable after conventional TMS treatment. Theta Burst Stimulation can keep away from that limit and is by and large viewed as the most significantly progressed of each of the various kinds of TMS. A couple of studies[25] have come out which have tried many sittings of TBS in a single day (e.g., 5 sittings each day over 6 days), which is called Accelerated TMS.[26] In any case, the jury is still out on whether this is pretty much powerful than the standard 6-week course.

Structural variation of figure-eight coils

Therefore, the development of the figure–eight coils, scientists have continually aimed to work on its performance. The focus of these efforts remains on the coil winding since it incredibly influences the performance of the coil. Throughout the long time, different alterations of the coil configuration and their subsequent benefits have been accounted for. The loop is bowed at an intense point at the middle between the left and right wings. The bending forces the coil to conform to the shape of the human head.

In addition, it brings about an increase depth electric fields in the brain. This loop attains a high electric–field intensity at the junction of the wings. Furthermore, the twisted shape causes a decrease in the current intensity in nearby areas. Hence, this coil upgrades the concentration of the induced electric field. Krasteva et al., 2002 stated that slinky coils, are one more variety of figure–eight coils. Due to the commitments of coils components having obtuse angle, the coil gives electric field distribution with an improved concentration.

When biphasic pulse currents are applied to the two coils with a stage shift of a quarter cycle, the resultant electric field pivots in the plane of coil. The excitability of brain tissues relies upon the direction of the electric field. Accordingly, the rotating electric fields give stable stimulating impacts no matter what the coils direction is. The electric power productivity of figure-eight coils can be enhanced by utilizing the core of the iron. Yamamoto et al., 2016 stated in the research that, this empowers the production of strong magnetic fields with little driving currents. modest driving flows. The development of power productivity is valuable for therapeutic uses that need frequent stimulations with frequencies more than 5 Hz. Nevertheless, the iron core ought to be made to limit eddy currents in the core.

Role of TMS in stroke

Stroke is a main root of disability, and the burden of stroke is commonly seen in elderly population having great incidence of ischemic stroke compared to younger population. For every succeeding 10 years after the age of 55 years, the stroke rate becomes more than doubles in both genders. Many patients are left completely disabled, with some secondary complications such as motor impairments, cognitive and language problems, and psychological problems.[27] Recent researchers found that more than one-third patients yet needed assistance for ADL and around one-third patients left disabled.[28]

Recovery is complicated after stroke. Many interventions have been evolved to help in recovery of impairment and related capabilities.[29] A non-invasive brain stimulation procedure namely transcranial magnetic stimulation (TMS) has been developed. This stimulation interacts with natural brain activity and impacts sensory and motor functions and higher-order cognitive functions. In the 1980s, TMS was established for the analysis of the functional state of the corticospinal pathway.[30] In the 1990s, technological advances allowed the transferring of rhythmic trains of magnetic pulses in a fast sequence with repetition rate of till 100-Hz, called as repetitive TMS (rTMS). It was seen that rTMS, interacts with cortical actions greater successfully than TMS.[31]

The commonest impairment caused by stroke is motor impairment, with loss of motor function or reduction in mobility.[18] Hence, the main focus of stroke rehabilitation is regaining impaired movement and the function related to motor learning. rTMS is a effective tool as it creates effects on cortical excitability, stroke damages the balance of transcallosal inhibitory pathways between primary motor areas in both hemispheres, the damaged hemisphere may be disrupted by infarction and asymmetric inhibition from healthy hemisphere. Hence, rTMS can restore the balance. According to the interhemispheric competition model, there are two types of procedures – down regulation of the excitability of the primary motor cortex in the healthy hemisphere with low-frequency stimulation, and up regulation of excitability in the affected hemisphere with high-frequency stimulation.[32]

Consecutive multi-session trials for acute and chronic stroke in children and adults have demonstrated the efficacy of the down regulation method. Contrastingly, the up-regulation approach hasn't been used very often, mostly because of safety worries because it was believed that high-frequency rTMS would raise the chance of seizures. The evidence supporting the safety and effectiveness of high frequency rTMS to the motor cortex of the damaged hemisphere, however, was recently evaluated by Corti et al.[33] The studies that were examined in this systematic review looked at how rTMS affected adult stroke survivors' upper-limb motor function and corticospinal pathway excitability at the same time. The scientists concluded that rTMS administered to the affected hemisphere is a safe procedure and may be a useful strategy for modifying brain activity and assisting with motor recovery following stroke. Additionally, some motor cortex researchers hypothesized that stimulating both areas would be the best course of action.[34]

Role of TMS in cerebral palsy

One of the most prevalent neurodevelopmental disorders in children is cerebral palsy (CP). It can result from a head injury during Labouré or after delivery, brain damage from an insult, or from a premature birth. Many speeches, visual, fine and gross motor function, and cognitive impairment issues are linked to it.[35] The purpose of our study was to determine the Effect of rTMS in the improvement of understanding and cognitive abilities by using magnetic stimulation in patients and analyzing changes in recorded Electroencephalogram (EEG) data after rTMS intervention, rTMS is one of the most recent technologies being used for deep brain stimulation.[36] Fast Fourier Transform (FFT) and Power Spectrum Density (PSD) calculations were made from the EEG signals for data processing.

For rTMS in this study, Neuro-MS. NET software and a therapeutic magnetic stimulator with a figure–eight-shaped angulated coil were used. The four Tesla magnetic field produced by the eight-shaped magnetic coil penetrates the skull up to six centimeters, triggering the neurons in the underlying brain area. One TMS pulse is used to evaluate MT, and the magnetic coil is placed on the appropriate dermatome to cause a twitch to appear in the targeted area. Once the degree of muscle twitching has been determined, the MT is fixed for the duration of treatment. Prior to performing the TMS therapy, the patient must be comfortable and relaxed, and certain important personal preparations must be made. This is done to ensure that the TMS effectively reaches the targeted stimulation area.

Cognitive impairment is a key feature in CP that restricts social and intellectual development. Performance in the areas of cognition and memory is correlated with the quantity of EEG power in the theta and alpha frequency range.[37] Small theta power but large alpha power in the EEG, according to Klimesch et al., 1996, implies good performance.[38] It was claimed that theta waves and the encoding of incoming information have a close link.[39] Givens et al. (1995) reported in one study that pharmacological interventions reduce theta wave activity, which inhibits learning.[40] Gamma waves may be responsible for the visual encoding process, which aids in learning through visual means, according to a further study by Gray et al. in 1987.[41] This study made a strong case for rTMS as a potent technique with the ability to neuromodulate brain signals. The rTMSs excitatory properties aid in the activation of neurons that created new neural connections in the cerebral cortex by boosting mental activity. The excitatory quality of rTMS aids in the activation of neurons that created new neural pathways in the cerebral cortex by elevating mental activity. Children with CP see an improvement in cognitive abilities because of these enhanced mental processes, which awaken dormant brain structures, enabling momentary connections through communication gaps called synapses and subsequently improving the overall neural link. Researchers discovered a substantial improvement in the cognitive domain with the use of rTMS in EG as compared to CG with significant P values for the EG in one of our earlier research projects of Gupta et al., 2017 on CP Patients. In a different study by Guse et al., 2010,[42] they discovered that there was a selective improvement in cognitive function after using high-frequency TMS stimulation over the left dorsolateral prefrontal cortex (DLPFC). They looked at how clinical state affected patients' cognitive outcomes as well as those of healthy volunteers. The findings of this study indicated that patients typically make more progress than healthy participants. Thirty people with spastic CP were recruited for the current investigation. 15 of the participants simply experienced ADLs (CG), whereas the remaining people (EG) received rTMS stimulation at a preset frequency of 10 Hz. The observations of their brainwave patterns and physiology revealed clear changes. Although epileptic discharge could be regarded one of the linked clinical features seen in CP patients, the temporal association between EEG recording and TMS shows no epileptic discharge during the TMS stimulation in each patient. It demonstrates lowered theta wave power peak values and increases alpha wave power peak values, which are associated with improved learning and the acquisition of new skills in children with CP. Statistical evaluation Theta wave peak frequency readings show a value of 15 Hz, however there is little variation in their strengths. The findings suggested that rTMS improves learning capacity in children with cerebral palsy, and that it will be a useful tool for memory enhancement in the future, particularly for treating neurodegenerative diseases in children and adults, slow learners, Alzheimer's patients, and patients with cognitive impairment. Better outcomes with rTMS are anticipated in the rehabilitation sector in the future for cases with cognitive impairment. Contrastingly, CG results show that daily living activities (ADLs) are not effective in improving cognitive function because they are routine tasks that are practiced frequently. Instead, they help relieve the mental stress of children who have difficulty performing ADLs during a specific time, which is a challenging job for them.

Role of TMS in Alzheimer's disease

The most prevalent kind of dementia among elderly people is Alzheimer's disease (AD), which is characterized by memory loss as its primary symptom. Loss of direction, abnormal conduct, a lack of desire, depression, and motor impairments are additional symptoms.[43] Early-stage therapies are now more crucial than ever because of the suffering of patients and the cost to society. The long-term prognosis is not improved by drugs, which have minimal effects.[44]

The pooled effect of rTMS with long-term treatment was modest, while the effect of short-term treatment was extremely little, according to the subgroup analysis of the number of sessions. As is common knowledge, rTMS is a painless, non-invasive device that can detect and control cortical excitability and functioning.[45] According to earlier research, individuals with neurodegenerative diseases, particularly AD and Parkinson's disease, benefited more significantly and for a longer period from repetitive and long-term stimulation.[46] This meta-analysis supports the earlier research' findings that longer stimulation times result in stronger rTMS effects.

According to a subgroup analysis, adding rTMS to prescription drugs or cognitive training (6 studies, 119 individuals each) did not improve results (8 studies, 151 people total). In contrast, Cheng et al.[47] in his meta-analysis studies on rTMS that combined with cognitive training rather than medications to enhance cognition was more effective at improving cognitive function in AD patients. Although our meta-analysis found that rTMS and cognitive training did not result in further cognitive improvement, we believe that rTMS and cognitive training (rTMS-COG) may result in higher progress. The fact that there were few participants, and the sample sizes of the groups were out of equilibrium—the largest group consisted of 30 participants, while the smallest group consisted of 8—could be one explanation for the outcome. According to Rabey et al.[48] considerable cognitive improvement was seen in AD patients who used rTMS-COG, indicating that rTMS-COG is a beneficial adjuvant therapy for AD patients. Additional research must be incorporated to more thoroughly validate this claim.

Subgroup Analysis was not possible because there was just one study[49] that directly compared low-frequency and high-frequency rTMS on cognition related to AD. High-frequency rTMS, as opposed to low-frequency rTMS, has been shown in multiple studies to be more effective than the latter in treating AD.[50] Therefore, picking the best rTMS frequency is essential.

There are a few restrictions on our current meta-analysis that need to be considered. First, our metanalysis's sample size and study count were both modest. Second, due to insufficient data, researchers were unable to examine the effect of duration even though we assessed the effectiveness of rTMS. Third, it was expected that there would be differences between research, and this discrepancy may have affected our findings. We chose the best RCTs based on strict inclusion and exclusion criteria to avoid this issue. Finally, researchers did not employ a comprehensive evaluation of cognitive performance in AD patients since we only used a few evaluation items to measure cognitive function.

In conclusion, our meta-analysis demonstrated that rTMS Therapy can considerably enhance cognitive function in AD patients. Long-term treatment and multiple site stimulation—most notably bilateral DLPFC stimulation are more efficient at enhancing AD-related cognitive performance. A more effective treatment site to enhance memory in AD may be some novel interventional targets, such as PC.

Role of TMS in dementia

Following Alzheimer's disease (AD), vascular dementia (VaD) is thought to be the second most prevalent type of dementia, with a prevalence of 2.6 percent in participants aged 1–65 in European studies.[51] The most prevalent type of cognitive disorder, however, is vascular cognitive impairment, which includes any degree of cognitive impairment from an impairment of a specialized cognitive function to VaD.[52] It affects about 5 percent of people over the age of 65. There is now consensus that VaD results from various vascular pathologies, including subcortical ischemic small-vessel disease, cortical infarcts, ischemic-hyperperfused lesions, or hemorrhagic lesions that can cause a variety of clinical phenotypes, contrary to the outdated concept of the multi-infarct dementia model.[53] VaD diagnosis and categorization have become more challenging and contentious because of this clinical and pathological variability.[54]

Studies on rTMS as a therapy for VaD are scarce currently, and its posited mechanisms are yet unknown. According to the Rektorova et al. randomized controlled pilot trial,[55] one session of high frequency rTMS applied over the left dorsolateral prefrontal cortex was able to increase executive performance while having no noticeable effects on any other cognitive abilities. However, only a small number of individuals with mild cognitive impairment but no dementia who were suffering from CVD were included in the study. Also examined for the treatment of elderly patients with vascular depression, which is known to be more frequently drug-resistant than early-onset depression,[56] were the efficacy and safety of high-frequency rTMS.

Recent research of rTMS in rat models of VaD found that the administration of either low-frequency or high-frequency stimulation improved both learning and memory abilities. The authors speculate that the biological effects of rTMS, which likely act by promoting the expression of brain-derived neurotrophic factor and other proteins, involved in neuronal cell protection, synaptic transmission, and brain plasticity, may be the cause of the restoration of cognitive functions.[57] Most TMS data show that the cortex of VaD patients is hyperexcitable, a characteristic also presents in AD. There is a dearth of literature on TMS in the early stages of VaD. The scant research on early AD and amnestic MCI cortical motor neuron excitability was not significantly altered.[58] Future research combining TMS with biomarkers and neuroimaging may be useful in identifying the brain regions at risk for dementia and in predicting the course of the disease.

As a result, even though there have been numerous neuroradiological reports on VaD, there have only been a small number of neurophysiological investigations that have looked at cortex excitability in patients who have had cognitive impairment due to vascular brain damage. All the data point to the fact that TMS is already a useful tool for researching the neurophysiological underpinnings of cognitive problems. To comprehend how subcortical vascular lesions affect cortical excitability and the function of multiple neurotransmitter involvement in VaD patients, further TMS studies are required.

The predementia stage has not yet been the subject of any investigations, and the relationship between cortical excitability and cognitive function has not been explored. There is very few research on how medications affect vascular dementia patients' cortical excitability. Studies using rTMS, however, can offer information for a therapeutic application aimed at enhancing cognitive function in VaD.

Role of TMS in autism

According to the Diagnostic and Statistical Manual of Mental Disorders, Fifth Edition (DSM-5), autism spectrum disorder (ASD) is a complex neurodevelopmental disorder characterized by persistent deficits in social interaction and communication as well as stereotyped behaviors, interests, and activities.[59] According to the most recent US Centers for Disease Control and Prevention data, 1 in 68 children now have ASD.[60] According to these findings, ASD is the most prevalent neurodevelopmental disease. The enormous social, psychological, and financial costs of ASD are as a result.

Aberrant neuroplasticity is one mechanism that has recently garnered a lot of evidence for its potential significance in the pathophysiology of ASD.[61] There are numerous lines of evidence, ranging from genetics,[62] animal models,[63] neuroimaging,[64] and brain stimulation.[65] Research has started to link ASD to abnormal neuroplasticity. TMS and rTMS are two examples of neuroscientific tools that have gained popularity as secure, non-invasive ways to examine abnormal neuroplasticity. The use of TMS/rTMS in the research of schizophrenia, depression, and Parkinson's illness is maybe a good illustration of this.[66] Beginning to be understood is the diagnostic and therapeutic potential of rTMS in ASD. In this article, we will briefly review evidence of abnormal neuroplasticity in autism spectrum disorders (ASD), recommend future directions in neuroplasticity assessment using rTMS, and talk about the potential of rTMS in correcting abnormal neuroplasticity in ASD.

For research and therapeutic applications, repetitive TMS, which comprises repetitive delivery of pulses (>1 Hz), is utilized to alter cortical activity. Neuroplasticity in humans has been increasingly studied with rTMS. The fundamental idea is that rTMS can alter brain activity in the targeted area over a period of time that may outlast the effects of stimulation.[67] According to current thinking, rTMS causes such long-lasting alterations in the brain by changing neuroplasticity mechanisms.[68]

Researchers now have the opportunity to create customized stimulation protocols that can modify neuroplasticity thanks to repetitive TMS, and these neuroplasticity-based brain stimulation therapies appear promising. Our team recently showed that the application of 1,500 pulses of high frequency (20 Hz) rTMS to the DLPFC can "normalize" working memory deficits in schizophrenia. This was done in a randomized double-blind sham-controlled study.[69] Enhancing neuroplasticity in the DLPFC is one potential mechanism for such improvement. To treat abnormal neuroplasticity in ASD, similar treatment approaches should be investigated.

In conclusion, evidence from human rTMS research and existing genetic and animal investigations of ASD has consistently pointed to abnormal neuroplasticity in the ASD brain. The precise etiopathological relationship between abnormal neuroplasticity in the brain and the onset of autism symptoms is still unclear at this time. However, the information that is now available suggests that abnormal neuroplasticity may play a significant role in the pathophysiology of ASD. Therefore, it can be hypothesized that stabilizing aberrant neuroplasticity may make it possible for ASD patients to achieve optimal social and cognitive performance.

Repetitive transcranial magnetic stimulation (rTMS) is an as of late evolved painless non –invasive brain stimulation technique for the treatment of neurological and psychiatric problems. It includes the use of changing magnetic field to the superficial layers of the cerebral cortex, which contains little electric currents, called as 'eddy current'. In this circumstances, cerebral cortex work as an secondary coil. TMS enjoys a benefit as it is concentrated and bypass the impedance of skull and superficial layers. Despite the fact that, its mechanism of action is still not revealed, current proof focuses toward its part in causing long term excitation and inhibition of neurons in specific cerebrum regions. Evidence in support of rTMS as therapeutic tool continues to grow, there is a need to develop standardized protocols for its organization. It does not cause any major reactions with rTMS, however its utilization is confined in those having magnetic implants or neurological or heart problems. Among all the indication of rTMS, the proof is more towards the treatment of refractory unipolar depression. Observations on electrical excitement of cerebral cortex began in 1874 in which contralateral motor feedback was evoked. In 1881, Faraday gave the principal on electro-magnetic induction. But Kolin et al. in 1959 accomplished nerve excitation by involving magnetic energy in frogs which established the groundwork for EM stimulation of brain tissue for diagnostic and therapeutic purposes. In last couple of years, rapid progress has been made in the development of coil shapes to concentrated magnetic field to attain better command over the spatial extent of excitation. While the old type of treatment took up to 37minutes per sittings, with high-frequency (HF) theta-burst stimulation the sittings may only last a few minutes. It is logical that treatment conventions will go through additional refinements in the coming years to make it more comfortable for patients.

Deep brain stimulation

Deep brain stimulation (DBS) involves implanting of electrode by mean of surgical process in the central areas of the brain. These electrodes produce electrical impulses for the regulation of abnormal impulses. In DBS stimulation device is controlled by a pacemaker[70] like device which is placed beneath the skin of chest and this wire connects the electrodes to the brain. DBS is usually used for the treatment of movement disorders such as Parkinson's disease, essential tremor,[71] dystonia, epilepsy and spasticity. It is one of the most established treatments now a day for the management of movement disorders such as Parkinson's disease. Parkinson's patient generally having very severe motor issues such as postural insufficiency, tremors, freezing movements and gait disorder DBS.

Mechanism of deep brain stimulation

At molecular level the sign and symptoms of parkinsonism results due to deficiency of dopamine.[72] According to neurophysiological point of view the motor symptoms seen due to defect in the basal ganglia[73] and segregation of thalamocortical circuit in the human brain. The dysfunction of thalamocortical circuit or any other tracks disrupts the thalamus, cortex and brainstem activities. DBS is the new approachable device which helps to free these down streaming.

Risk factors and side effects

DBS also have many complications and risk factors due to surgical intervention such as misplacement of electrodes, excessive bleeding in the brain, stroke, infection, nausea breathing problem and seizure. Its side effects are also associated due to surgical intervention such as headache, confusion, difficulty in maintain concentration, and temporary pain and swelling at the implantation side.

DBS in various neurological conditions

DBS is very effective in the various movement disorders conditions such as Parkinson's (PD), dystonia, Huntington's disease. DBS generally consist of three components such as implanted pulse generator (IPG), the lead and an extension. The implanted pulse generator sends the electrical impulses to the brain that interfere with neural activities at the targeted site.[74] The lead is connected with IPG by extension.

Parkinson's disease

In recent research it is found[75] DBS stimulates the neurogenesis and release neuroprotective chemicals in Parkinson disease. It enhances the activity of caudate nucleus and release dopamine. According to may literature DBS potentially used as an important adjunctive treatment for PD.[76] It's also suggested as potential treatment for the cognitive enhancement.[77] According to recent literature it is found that DBS have potential to manage many symptoms of PD where medication[78] is not effective so much even medication have some side effect. In Parkinson disease motor symptoms are the targeted objectives in DBS such as slow motion, stiffness, tremors and abnormal movements.[79] Apart from benefits of DBS in Parkinson's Patients its also have some side effect such as bleeding, hemorrhage, infection, sometime inaccurate placement of electrodes, seizures, confusions, pain and swelling.

Dystonia

Dystonia is a neurological disorder that results excessive uncontrolled muscle contraction, which results abnormal muscle movement and body alignment. It can affect region any region of the body includes the face, eye movement or any part of the body it can be seen depends upon which lesion of the brain is involved. Till now the main mechanism of action for improvement of dystonia is still not clear but cortical hyper-excitability in dystonia can cause abnormalities in the sensorimotor system.[80] With the help of DBS, we can increase intercortical inhibition and reduce excessive cortical plasticity.

Huntington's disease

Huntingtin's disease (HD) is an autosomal –dominant neurodegenerative disorder which have very less results with therapeutic intervention as well as with medication. The DBS helps in modulating the brain activities which is very difficult to achieves with the help of medication or any kind of surgical intervention. In Huntington's disease frontal area of brain[81] can be affected by DBS by increasing flow of dopamine to various part of the brain including dorsal striatum that generates the passage that is very effective in treatment of Huntington's disease.[82]

DBS in Alzheimer's disease

The underlying principle and mechanisms of DBS is not fully understood but it directly changes the brain activity which is approved by Food Drug Administration for Parkinson since 1997. As per the recent research it found that DBS have potential to restore circuit dysfunction in Alzheimer's disease and it also reverse the slow the process of underlying disease pathophysiology.

DBS is a neurosurgical technique which regulates the neuronal activities by internal pulse generator to the targeted area of brain with electrodes. DBS is blindly using for the treatment purpose in many psychiatric and neurological disease, but the mechanism of action is still unclear. DBS has shown a certain affective result in the stimulation parameters which determining the optimal stimulation in Alzheimer's disease. It enhances the cortical excitability with high frequency stimulation.[83] It has potential to change synaptic strength, which helps in learning and registration of memory.[84] It regulates the neuronal activities and increase neuroplasticity in cognitively decline patients.

Discussion

Brain stimulation in rehabilitation helps to increase the rate of progression in the patients. According to Majed Aldehri 2018 it is found in rodents' models DBS is most beneficial for early stage of Alzheimer's disease.[85] It helps to do long term structural neuroplasticity such as hippocampus enlargement and increase discharge of neurotransmitter release. In one of the studies by Jorden Lam 2021, it is found that neural circuit dysfunction[86] at functional and structural level can be improve and it indicate the potential restoration of circuit function.

Conclusion

Rehabilitation is one of the most import parts of any kindly treatment where it's surgical or medical. There are various traditional rehabilitation methods are available such as physical rehabilitation or occupational rehabilitation. Now a days to increase the rate of progress of any kindly of treatment recent advancement is required. In field of recent advancement repetitive transcranial magnetic stimulation is the one of most promising intervention for the brain stimulation. It is approved by food drug administration of US for many neurological as well as psychiatric condition which involve deterioration of movement disorders such as cerebral palsy, Parkinson's disease. It helps in neuroplasticity by excitation and inhibition of neural fibres with the stimulation.

Brain stimulation in any neurological condition helps to enhances the rate of progress for any neuro-muscular disease. Now a day there is recent advancement in field of research and clinical practices which make more promising treatment such as DBS, TMS and many more brain stimulation biomedical devices. These biomedical devices is also proven for the use on human body for neuro muscular disorders where medication have no such role accept to improves the secondary symptoms. TMS, DBS is the promising tool now a day for brain stimulation rehabilitation which is approved from food drug administration (FDA) in many clinical conditions from U.S. It is use for clinical, diagnostic as well as research purpose in all over the world. It is safe and well tolerated by most of the patient will mildly side effect such as mild headache. It has long lasting effect and it increase the rate of progress of other intervention together. There is need to develop well standardized protocols for its application and develop affordable therapeutic tool for societal impact.

References

1. Rastogi P, Lee EG, Hadiani RL. Transcranial magnetic stimulation: development of a novel deep brain triple halo coil. *IEEE Magnet Let*. 2019;10:1–5.
2. Rostogi P. *Novel Coil Designs For Different Neurological Disorders in Transcranial Magnetic Stimulation*. Ames: Iowa State University; 2019.
3. Dionísio A, Duarte IC, Patrício M, Castelo-Branco M. The use of repetitive transcranial magnetic stimulation for stroke rehabilitation: a systematic review. *J Stroke Cerebrovasc Dis*. 2018;27:1–31.
4. Deng Z-D, Lisanb SH, Peterchev AV. Electic field depth focality tradeoff in transcranial magnetic stimulation: stimulation comparison of 50mcoil designs. *Brain Stimul*. 2013;6:1–13.
5. Roth Y Zangen A, Hallett M. A coil design for transcranial magnetic stimulation of deep brain regions. *J Clin Neurophysiol*. 2002;19:361–370.
6. Atorre A, Rocchi L, Berardelli A, Bhatia KP, Rothwell JC. The interindividual variability of transcranial magnetic stimulation effects: implications for diagnostic use in movement disorders. *Mov Disord*. 2019;34:936–949.
7. França C, de Andrade DC, Teixeira MJ, Galhardoni R, Silva V, Barbosa ER, et al. Effects of cerebellar neuromodulation in movement disorders: a systematic review. *Brain Stimul*. 2018;11:249–260.
8. Sakai K, Hikosaka O, Miyauchi S, Takino R, Sasaki Y, Pütz B. Transition of brain activation from frontal to parietal areas in visuomotor sequence learning. *J Neurosci*. 1998;18:1827–1840.
9. Voigt J, Carpenter L, Leuchter A. cost effectiveness analysis comprising transcranial magnetic stimulation toantideprressant medicationsafter a frist treatment failure for major depressive disorder in newly diagnosed patient –a lifetime analysis. *PLoS One*. 2017.
10. Wang Y, Zhan G, Cai Z. Vagus nerve stimulationin brian disease. *Therapeutic application and biological mechanism*. 2021;127:37–53.
11. George MS, Sackeim HA, Rush AJ, et al. Vagus nerve stimulation: a new tool for brain research and therapy. *Biol Psychiatry*. 2000;47(4):287–295.
12. Lanska DJ. J.L. Corning and vagal nerve stimulation for seizures in the 1880s. *Neurology*. 2002;58(3):452–459.
13. Bailey P, Bremer F. A Sensory Cortical Representation of the Vagus Nerve: with a Note on the Effects of Low Blood Pressure on the Cortical Electrogram. *J Neurophysiol*. 1938;1(5):405–412.
14. Helmers SL, Duh MS, Guérin A, et al. Clinical and economic impact of vagus nerve stimulation therapy in patients with drug-resistant epilepsy. *Epilepsy & Behavior: E&B*. 2011;22(2):370–375.
15. Revesz D all. Complication and safety of vagus nerve stimulation; 25 years of experiences at a single centre. *J Neurosurg Pediatr*. 2016;18(1):97–104. doi:10.3171/2016.1.PEDS15534.

16. Ugawa Y, Genba-Shimizu K, Kanazawa I. Suppression of motor cortical excitability by electrical stimulation over the cerebellum in Fisher's syndrome. *J Neurol Neurosurg Psychiatry*. 1994;57:1275–1276.
17. Ugawa Y, Hanajima R, Kanazawa I. Motor cortex inhibition in patients with ataxia. *Electroencephalogr Clin Neurophysiol*. 1994;93:225–229.
18. Hanajaima R, Tsutsumi R, Shirota Y, Shimizu T, Tanaka N, Ugawa Y. Cerebellar dysfunction in essential tremor. *Mov Disord*. 2016;31:1230–1234.
19. Boggio PS, Ferrucci R, Rigonatti SP, et al. *Neurol Sci*. 2006;249(1):31–38.
20. Cohen Kadosh R, Elliott P. Neuroscience: brain stimulation has a long history. *Nature*. 2013;500:529.
21. Kolin A, Brill NQ, Broberg PJ. Stimulation of irritable tissues by means of an alternating magnetic field. *Proc Soc Exp Biol Med*. 1959;102:251–253.
22. Blumberger DM, Vila-Rodriguez F, Thorpe KE, Feffer K, Noda Y, Giacobbe P, et al. Effectiveness of theta burst versus high-frequency repetitive transcranial magnetic stimulation in patients with depression (THREE-D): a randomised non-inferiority trial. *Lancet*. 2018;391:1683–1692.
23. Chung SW, Hoy KE, Fizgerald PB. Theta-burst Stimulation: a New Form of TMS Treatment for Depression? *Depress Anxiety*. 2015;32:182–192.
24. Hong Y, Wu S, Pedapti E, et al. Safety and tolerability of theta burst stimulation vs. single and paired pulse transcranial magnetic stimulation: a comparative study of 165 pediatric subjects. *Front Hum Neurosci*. 2015 Feb 03;09:1–5.
25. Bulteau S, Sébille V, Fayet G, et al. *Trials*. 2017 Jan 13;18(1):17. doi:10.1186/s13063-016-1764-8.
26. Prasser J, Schecklmann M, Poeppl TB, et al. *World J Biol Psychiatry*. 2015 Jan;16(1):57–65.
27. Paul SL, Srikanth VK, Thrift AG. The large and Growing burden of stroke. *Curr Drug Targets*. 2007;8:786–793.
28. Feigin VL, Barker-Collo S, McNaughton H, Brown P, Kerse N. Long-term neuropsychological and functional outcomes in stroke survivors: current evidence and perspectives for new research. *Int J Stroke*. 2008;3:33–40.
29. Langhorne P, Bernhardt J, Kwakkel G. Stroke rehabilitation. *Lancet*. 2011;377:1693–1702.
30. Barker AT, Jalinous R, Freeston IL. Non-invasive Magnetic stimulation of human motor cortex. *Lancet*. 1985;i:1106–1107.
31. Hummel FC, Cohen LG. Non-invasive brain stimulation: a new strategy to improve neurorehabilitation after stroke? *Lancet Neurol*. 2006;5:708–712.
32. Fitzgerald PB, Fountain S, Daskalakis ZJ. A comprehensive review of the effects of rTMS on motor cortical excitability and inhibition. *Clin Neurophysiol*. 2006;117:2584–2596.
33. Corti M, Patten C, Triggs W. Repetitive transcranial Magnetic stimulation of motor cortex after stroke: a Focused review. *Am J Phys Med Rehabil*. 2012;91:254–270.
34. Takeuchi N, Tada T, Toshima M, Matsuo Y, Ikoma K. Repetitive transcranial magnetic stimulation over Bilateral hemispheres enhances motor function and Training effect of paretic hand in patients after stroke. *J Rehabil Med*. 2009;41:1049–1054.
35. Gupta M, Bhatia D, Rajak BL. Study of available intervention technique To improve cognitive function in cerebral palsy patients. *Curr Neurobiol*. 2017;8:51–9.
36. Gupta M, Bhatia D. Effect of repetitive transcranial magnetic stimulation On cognition in spastic cerebral palsy children. *J Neurol Disord*. 2017;5:1–2.
37. Klimesch W. EEG alpha and theta oscillations reflect cognitive and Memory performance: a review and analysis. *Brain Res Brain Res Rev*. 1999;29:169–95.
38. Klimesch W. Memory processes brain oscillation and EEG synchronization. *Int J Psychophysiol*. 1996;24:61–100.
39. Miller R. *Cortico–Hippocampal Interplay and the Representation of Contexts in the Brain*. Berlin: Springer; 1991.

40. Givens B, Olton D. Bidirectional modulation of scopolamine in duced Working memory impaired by muscarinic by muscarinic activation of the Medical Septal area. *Neurobiol Learn Mem.* 1995;63:269–76.
41. Grey C, Singer W. Stimulus dependent neuronal oscillations in the cat Visual cortex area 17. *Neuroscience.* 1987;22(Suppl):434.
42. Guse B, Falkai P, Wobrock T. Cognitive effects of high–frequency Repetitive transcranial magnetic stimulation: a systematic review. *J Neural Transm (Vienna).* 2010;117:105–22.
43. Brini S, Sohrabi HR, Peiffer JJ, Karrasch M, Hamalainen H, Martins RN, et al. Physical activity in preventing Alzheimer's disease and cognitive decline: a narrative review. *Sports Med.* 2018;48(1):29–44.
44. Bachurin SO, Bovina EV, Ustyugov AA. Drugs in clinical trials for Alzheimer's Disease: the major trends. *Med Res Rev.* 2017;37(5):1186–1225.
45. Vucic S, Kiernan MC. Transcranial magnetic stimulation for the assessment of Neurodegenerative disease. *Neurotherapeutics.* 2017;14(1):91–106.
46. Anderkova L, Rektorova I. Cognitive effects of repetitive transcranial magnetic Stimulation in patients with neurodegenerative diseases – clinician's perspective. *J Neurol Sci.* 2014;339(1–2):15–25.
47. Cheng C, Wong C, Lee K, Chan A, Yeung J, Chan W. Effects of repetitive transcranial magnetic stimulation on improvement of cognition in elderly patients with Cognitive impairment: a systematic review and meta-analysis. *Int J Geriatr Psychiatry.* 2018;33(1):e1–e13.
48. Rabey JM, Dobronevsky E. Repetitive transcranial magnetic stimulation (rTMS) Combined with cognitive training is a safe and effective modality for the treatment Of Alzheimer's disease: clinical experience. *J Neural Transm.* 2016;123(12):1449–1455.
49. Ahmed MA, Darwish ES, Khedr EM, El Serogy YM, Ali AM. Effects of low Versus high frequencies of repetitive transcranial magnetic stimulation on cognitive Function and cortical excitability in Alzheimer's dementia. *J Neurol.* 2012;259(1):83–92.
50. Nardone R, Tezzon F, Holler Y, Golaszewski S, Trinka E, Brigo F. Transcranial Magnetic stimulation (TMS)/repetitive TMS in mild cognitive impairment and Alzheimer's disease. *Acta Neurol Scand.* 2014;129(6):351–366.
51. Fratiglioni L, Laurer LJ, Andersen K, et al. Incidence of dementia and major subtypes in Europe: a collaborative study of population Base cohorts. *Neurology.* 2000;54:S10–S15.
52. Hachinski V. Vascular dementia: a radical Redefinition. *Dementia.* 1994;5:130–132.
53. Román GC, Erkinjuntti T, Wallin A, Pantoni L, Chui HC. Subcortical ischaemic vascular Dementia. *Lancet Neurol.* 2002;1:426–436.
54. Román GC, Tatemichi TK, Erkinjuntti T, et al. Vascular dementia: diagnostic criteria for research studies: report of the NINDS-AIREN International Workshop. *Neurology.* 1993;43:250–260.
55. Rektorova I, Megova S, Bares M, Rektor I. Cognitive functioning after repetitive transcranial magnetic stimulation in patients With cerebrovascular disease without dementia: a pilot study of seven patients. *J Neurol Sci.* 2005;229–230:157–161.
56. Fabre I, Galinowski A, Oppenheim C, et al. Antidepressant efficacy and Cognitive effects of repetitive transcranial Magnetic stimulation in vascular depression: an open trial. *Int J Geriatr Psychiatry.* 2004;19:833–842.
57. Wang F, Geng X, Tao HY, Cheng Y. The restoration after repetitive transcranial magnetic stimulation treatment on cognitive Ability of vascular dementia rats and its impacts on synaptic plasticity in hippocampal CA1 area. *J Mol Neurosci.* 2010;41:145–155.
58. Sakuma K, Murakami T, Nakashima K. Short latency afferent inhibition is not impaired in mild cognitive impairment. *Clin Neurophysiol.* 2007;118:1460–1463.
59. American Psychiatric Association. *Diagnostic and Statistical Manual of Mental Disorders 5th Edition (DSM-5).* Washington, DC: American Psychiatric Association; 2013.

60. Baio J. Prevalence of autism spectrum disorder among children aged 8 Years-autism and developmental disabilities monitoring network, 11 sites, United States, 2010. *MMWR Surveill Summ.* 2014;63:1–21.

61. Markram K, Markram H. The intense world theory – a unifying theory of the Neurobiology of autism. *Front Hum Neurosci.* 2010;21(4):224. doi:10.3389/Fnhum.2010.00224.

62. Dolen G, Bear MF. Fragile X syndrome and autism: from disease model To therapeutic targets. *J Neurodev Disord.* 2009;1:133–140. doi:10.1007/S11689-009-9015-x.

63. Markram K, Rinaldi T, LaMendola D, Sandi C, Markram H. Abnormal Fear conditioning and amygdala processing in an animal model of autism. *Neuropsychopharmacology.* 2008;33:901–912. doi:10.1038/sj.npp.1301453.

64. Courchesne E, Campbell K, Solso S. Brain growth across the life span in Autism: age-specific changes in anatomical pathology. *Brain Res.* 2011;22(1380):138–145. doi:10.1016/j.brainres.2010.09.101.

65. Jung NH, Janzarik WG, Delvendahl I, Münchau A, Biscaldi M, Mainberger F, et al. Impaired induction of long-term potentiation-like plasticity in patients With high-functioning autism and Asperger syndrome. *Dev Med Child Neurol.* 2013;55(1):83–89. doi:10.1111/dmcn.12012.

66. Rajji TK, Rogasch NC, Daskalakis ZJ, Fitzgerald PB. Neuroplasticity-based Brain stimulation interventions in the study and treatment of schizophrenia: a Review. *Can J Psychiatry.* 2013;58(2):93–98.

67. Pascual-Leone A, Freitas C, Oberman L, Horvath JC, Halko M, Eldaief M, et al. Characterizing brain cortical plasticity and network dynamics across the Age-span in health and disease with TMS-EEG and TMS-fMRI. *Brain Topogr.* 2011;24:302–315. doi:10.1007/s10548-011-0196-8.

68. Ziemann U, Paulus W, Nitsche MA, Pascual-Leone A, Byblow WD, Berardelli A, et al. Consensus: motor cortex plasticity protocols. *Brain Stimul.* 2008;1:164–182. doi:10.1016/j.brs.2008.06.006.

69. Barr MS, Farzan F, Rajji TK, Voineskos AN, Blumberger DM, Arenovich T, et al. Can repetitive magnetic stimulation improve cognition in schizophrenia? Pilot data from a randomized controlled trial. *Biol Psychiatry.* 2013;73(6):510–517. doi:10.1016/j.biopsych.2012.08.020.

70. Benabid AL, Pollak P, Louveau A, Henry S, de Rougemont J. Combined (thalamotomy and stimulation) stereotactic surgery of the VIM thalamic nucleus for bilateral Parkinson disease. *Appl Neurophysiol.* 1987;50:344–346.

71. Flora ED, Perera CL, Cameron AL, Maddern GJ. Deep brain stimulation for essential tremor: a systematic review. *Mov Disord.* 2010;25:1550–1559.

72. Cardoso SM, Moreira PI, Agostinho P, Pereira C, Oliveira CR. Neurodegenerative pathways in Parkinson's disease: therapeutic strategies, Current drug targets. *CNS Neurol Disord.* 2005;4:405–419.

73. DeLong MR, Wichmann T. Basal Ganglia Circuits as Targets for Neuromodulation in Parkinson Disease. *JAMA Neurol.* 2015;72:1354–1360.

74. Volkman J, Herzog J, Kopper F. Introduction to the programming of deep brain stimulation. *Mov Disord.* 2002;17(Suppl 3):S181–S187.

75. Wichmann T, Dostrovsky JO. Pathological basal ganglia activity in movement disorder. *Neuroscience.* 2011;198:232–244.

76. Habib S, Haid U, Jamil A, Zainab AZ Yousif T transcranial magnetic stimulation as a therapeutic option for neurologic and psychiatric illness: cureus (2018)

77. Khedr E, Mohamed KO, Ali A, Hasan A. The effect of repetitive transcranial magnetic stimulation on cognitive impairment in Parkinson's disease wit dementia: pilot study. *Restor Neural Neursci.* 2020;38:55–66.

78. Gildenberg PhL. Evolution of Neuromodulation. *Stereotact Funct Neurosurg.* 2005;83(2–3):71–79.
79. Caballol N, Martí MJ, Tolosa E. Cognitive dysfunction and dementia in Parkinson disease. *Mov Disord.* September 2007;22(Suppl 17):S358–S366.
80. Lozeron P, Poujois A, Richard A, Masmoudi S, et al. Contribution of TMS and rTMS in the understanding of the pathophysiology and in the treatment of dystonia. *Front Neural Circuits.* 2016.
81. Accolla EA, Aust S, Merkl A. Deep brian stimulation of the posterior gyrus region for treatment resistant depression. *J Affect Disord.* 2016;194:34–37.
82. Medina F, Tunez I. Huntingtons disease the value of transcranial magnetic stimulation. *Curr Med Chem.* 2010;17:2482–2491.
83. Weiler M, Stieger KC, Long JM, Rapp PR. Transcranial magnetic stimulation in Alzheimers disese : are we ready. *eNeuro.* 2020.
84. Guerra A, Assenza F, Bressi F, Scrascia F, Duca MD, et al. transcranial magnetic stimulation studies in Alzheimers disease. *Int J Alzheimers Dis.* 2011.
85. Mizadeh Z, Bari A. The rationale for deep brain stimulation in alzheimers disease. *Neural transm (vienna I).* 2016;123(7):775–783.
86. Lam J, Lee DJ. Response to letter to editor regarding Deep brain stimulation for alzheimers disease trackling circuit dysfunction. *Neuromodulation.* 2021;24(7):1291–1292.

CHAPTER 5

Advanced robotic rehabilitation

Introduction

Most neurological diseases influence the motor system of the body, causing handicaps incongruent with the individual's ordinary social capacities. A portion of these have a known, clear etiology and treatment; for other people, we are just toward the start of the street, considering here both pathogenesis and treatment. At the point when the motor system is involved, and similarly so in the event that the pyramidal or different parts are involved, a key objective is to quickly overcome the deficit. Standardized clinical conventions help to a point, yet physiotherapy-based recuperation plays likewise a significant part, irrespective of the location of the lesion. For instance, stroke causes central motor neuron diseases; motor neuron disease influence both central and peripheral fragments of the motor pathways; movement disorder, specifically in Parkinson's, impact modulation of the motor activity.[1] Restoration programs are intended to invigorate the plasticity of the involved pathways, supporting the recovery of motor capacities.[2] Regardless of whether regions of the motor cortex are interrupted by a vascular lesion stroke,[3] or by neurodegeneration, as in motor neuron diseases,[4] physiotherapy impacts the motor capacities of the patient. For stroke, a reasonable, quantifiable impact is estimated and the methodology turned into a standard for recovery,[5] both with classical physiotherapy[6] and with robotic-assisted rehabilitation.[7] Motor neuron disease advances, irrespective of the restorative methodology. physiotherapy, fundamentally assistive strategies, along with robotic rehabilitation measures,[8] appear to be advantageous in decreasing the speed of degradation.[9] Parkinson's patients benefit from multidisciplinary restoration,[10] physiotherapy being just a single methodology.[11] Gait, balance, and standing are well improved by physiotherapy recuperation,[12] with robotic methodologies being likewise widespread[13] for similar purposes.[14] Rehabilitation conventions are accessible[15] and may be carried out both by human therapists[16] or robotic devices.[17]

A successful physiotherapy program requires exercises a few times each day, a requesting task for the medical care provider. The demographic advancement of the population forces a change of the rehabilitation protocol

Modern Intervention Tools for Rehabilitation.
DOI: https://doi.org/10.1016/B978-0-323-99124-7.00005-5

worldview; later on, the patient/therapist proportion will turn out to be much more weakened, with the maturing of the population worldwide. It is especially essential to give solutions for the physiotherapist to have the option to work with numerous patients at a similar time. Since physical therapy regularly utilizes generalized, repetitive movements, the treatment may likewise be executed with the assistance of robots. A physiotherapist may just start, program, and manage such gadgets. One advisor can complete a few programs, associatively facilitating the increased lack of medical services providers. A rehabilitation robot, which is a robot that straightforwardly serves people, has broad application possibilities in rehabilitation treatment with high expert necessities. Thusly, it is vital to foster progressed rehabilitation robots.

Robotic rehabilitation

Each year, some 700,000 people in the United States have difficulty moving their upper extremities after having a stroke.[18] Treatment for movement impairments usually involves rigorous hands–on rehabilitation therapy. Patients only receive a little amount of this therapy because it is pricey and labour-intensive. Numerous research teams have created robotic devices to automate tedious portions of hands–on stroke therapy in order to solve this issue. With these devices, first clinical findings have been encouraging.[19] There is considerable agreement, though, that more advanced robotic technology would probably produce superior clinical outcomes. The creation of a tool that can apply a broad range of forces with strong dynamic bandwidth and securely help in fully naturalistic arm movements is of particular importance. A wide variety of therapeutic interactions and movement exercises could be used with such a gadget. Due to their high power-to-weight ratio, pneumatic actuators may be able to aid with the difficulty of lightweight, large force robots. Additionally compliant by nature, pneumatic actuators add an additional layer of security. Since they are challenging to control, they have not been frequently employed for robotic rehabilitation. The findings show that pneumatic actuation can be used to operate a robotic treatment device. The advantage of Pneumatic Actuators is their high power-to-weight ratio. In order to accomplish a typical goal in rehabilitation exercise, the apparatus presented here can raise a huge human arm and position it across a vast workspace. The device only has four degrees of freedom (DOFs), hence it can only move in a somewhat natural way for functional purposes. Because pneumatic actuators are often lightweight, adding additional degrees of freedom (such as shoulder internal/external rotation) is feasible. The fact that pneumatic actuators are more difficult to regulate than electric motors serves as a barrier to their use. This study demonstrates how a Kalman filter and inexpensive sensors may deliver the reliable sensing required for pneumatic control. This paper also offers a model of a low-cost servo valve that has been experimentally identified, enabling the robot to control force as requested. A brand-new, high-level controller was created to let the operator of the device move the arm to the intended target, but only when necessary. In order to preserve a compliant "feel" to the assistance while permitting accurate positioning, the controller develops an internal model of the static forces needed to keep the user's arm at various workspace positions. The fact that the robot can't be made rigid or have a very high force or position tracking bandwidth may be a drawback of the pneumatic technique used here. But the stiffness and bandwidth attained are comparable to the actual human arm. A pneumatic device like the one described here should be sufficient for optimal Rehabilitation results if suitable high level control strategies are used, just as the arm of a human therapist is sufficient to generate those results.

Our research on touch sensors was able to learn a lot from this work. First, by hermetically covering them with a silicon rubber film, we demonstrated the viability of converting commercial MEMS barometer sensors into force detectors.[20] By utilising the adaptability of the Kapton-based PCB, we also showed that such sensors may be plugged into a wearable thimble device, expanding the utility of earlier devices and enabling them to conform to the curved surface of the finger. To the best of our knowledge, this technique has never before been used to a data glove. Second, by positioning several sensors on the central, lateral, and medial aspects of the fingertip, tactile spatial dispersion was made possible. Because of this, the thimble system was able to provide both the points of contact and the overall interaction force. Thirdly, we were able to prevent signal cross-talk by wiring each single sensor into a standalone device. This problem was shown to be essential, calling for extra and sophisticated hardware solutions to reduce distortion effects on the output signals in matrix-based piezo-resistive tactile sensors.[21] Fourth, the device's compactness makes it exceedingly wearable and unobtrusive, which significantly lessens tactile impairment. When using complicated and taxing force transducers, this was demonstrated to be a problem.[22] The thimble device's performance enabled for the measurement of sensing forces at the level of the fingertip with a spatial resolution of approximately 10 and 5 mm in the central and lateral/medial aspects, respectively. With a fitting error of around 0.04 N and in accordance with,[20] the calibration technique produced an accurate linear mapping of the pressure/force relationship in the range of 0.035–3.74 N. The reduced force working range can be attributed to the accuracy result being better than other results reported in the literature.[23] Utilizing a commercial force/torque sensor, an independent test of the calibration quality was conducted (Nano17 – ATI Industrial Automation, Apex, USA). Interestingly, the created sensor and the ATI sensor had a good agreement on the force measurements that were done. The preliminary Grasp analysis found some diversity in the contact points for manipulation tasks. Greater than the test subjects. This may point to various gripping techniques, but it also shares traits like the prominence of the thumb and the lateralization of thumb contact regardless of the object's weight. The thimble device demonstrated the medium's less significant effect in compression pressures applied to the thumb and index finger.

The use of MEMS barometer sensors offers some advantages over existing force transducer technologies.[24] They have a simple communication interface and are commercial off-the-shelf technologies, which reduces the need for specialised electrical and mechanical knowledge. Second, the low cost of each sensor makes it theoretically possible to incorporate a sizable number of them into a data glove. Thirdly, the sensor's ability to simultaneously measure force and temperature is made possible by the fact that it can detect the ambient temperature. This might be an effective way for robotics to simulate the skin's real sensitivity. The approach here proposed uses isolated thimbles covering only the finger distal phalanges in contrast to glove-based solutions described in the literature.[25] This method maximises finger mobility and reduces burden in applications where only the fingertips are used.[26] It also avoids the need for sensorized gloves. Additionally, the system is simple to disassemble, the latex thimble size may be chosen with increasing adaptability based on finger size, and the same device can be used for both hands. Recent reports in the literature[23] describe a Thimble-based system called Thimble Sense that makes use of the Force/torque ATI sensor. The cost of a five-finger tactile system with around 30 sensors is less than about 500 times that of the Thimble Sense with all of the Five finger sensible, even if the MEMS barometer sensor can only sense orthogonal force.

Evaluation in the field of rehabilitation

The purpose of this study was to use a MEMS to estimate TEE during ADL while accounting for gender and HR. Based on entire body kinematics, a Wtot estimate was produced. By comparing them to a commonly used musculoskeletal model in the AMS setting, we validated our calculations of Wtot for sedentary activities and dynamic activities and discovered a good level of consistency. Wtot, HR, and gender were used as inputs in the construction of a straightforward linear regression model to estimate rEE. Our results shown that with our approach, even a straightforward regression model could accurately predict rEE in dynamic activities while only providing a passable forecast for rEE in static activities. Additionally, compared to dynamic activities, the relative LOA for sedentary activities appeared to be much larger. This supports our findings that sedentary activities have a lower quality of prediction, which has also been noted in prior studies.[27] The energy expenditures of maintaining a static posture are weakly re-Flected in the MEMS output, making it difficult to estimate energy consumption using MEMS during sedentary activities.[28] As the mean of TEE rises, a rising trend for sedentary activities. This shows that rEE is slightly underestimated by the sedentary model for inactive pursuits that are expensive despite being overestimated for low-cost sedentary activities, it may show how overlooked postural energy costs have a greater impact on low-cost activities.

The model's bias in dynamic activities indicates a little overestimation; this can be the result of the soft tissue's overlooked elastic energy being released. The dissipation of elastic energy can explain some of the mechanical alterations that have been seen. Balance of energy, although this may not always translate into intake of oxygen.[29] The Wtot calculations were validated, and the results showed low bias and a downward trend when the mean increased between the AMS and the MEMS. Additionally, only one sample was discovered to fall below the 95 percent agreement level. Again, it is caused by an increased contribution from the elastic energy release at faster running velocities.[30] Therefore, at large levels of energy turnover, our technique overestimates the cost of energy. While the bias confidence interval in absolute Bland-Altman plots does overlap zero, the bias confidence interval in the Bland-Altman plots for the relative differences between AMS and MEMS do not. This is a result of the recorded samples being overweighed after being adjusted to their mean values while seated as can be observed.

Our findings thus indicate a better TEE estimation for dynamic activities but not for sedentary ones. The studies' use of a single regression Eq. to predict both sedentary and dynamic activities may account for the improved fit for dynamic activities. Altini et al.'s findings[28] showing activity-specific models increased the TEE estimation accuracy lend credence to this. However, collecting the data from the inertial sensor to determine the type of activity and then applying various models is technically achievable.[31] It's hardly shocking that people who engage In sedentary activities are less fit. It is harder to estimate TEE when engaging in sedentary activities. MEMS estimation against dynamic activities[32] due to a small Wtot fluctuation range. Aside from that, it excludes the energy required to maintain a still position.[33] Although, we still manage to achieve a more reasonable fit for the sedentary model than models that have already been published, both static and dynamic activities.

Lower limb rehabilitation

Improving walking capacity is a significant objective of restoration and an essential concern regarding social and professional reintegration for an individual with neurologic-related gait impairment. Robots for lower limb gait recovery have been designed primarily to robotize repetitive labor-intensive training and to help therapists[34] and patients during different phases of neurorehabilitation. Part of this last mandate includes giving the most important kind of exercises, not simply in the number of repetitions, security, and motivation under which they are performed yet in addition as far as exercises that find some kind of harmony among accuracy and variability of movement, proper and unequivocally adjustable degrees of resistance, and ideal unweighting furthermore, help with performing the movements. The ideal qualities for these parameters are not known right now; notwithstanding, the training ideal models have their foundations in hypotheses of motor learning. motor learning mirrors neural explicitness of training since motor expertise procurement includes the combination of the sensory and motor data that happens during practice and, eventually, prompts an example that outcomes in accurate, reliable, and skilful movements. rehabilitation based on ideas of repetitive, intensive, task-oriented training has been demonstrated to be compelling. The restorative objectives of this approach are to accomplish restoration and recuperation of walking by harnessing the inborn capacities of the spinal and supraspinal locomotor centers.[35] Treadmill training with partial body weight support includes supporting a portion of the patient's weight over a mechanized treadmill while clinicians utilize manual help strategies to create stepping movements. The strategy plans to re-establish a typical physiologic walk design, with consideration regarding the ideal kinematic and temporal aspects of gait.[36]

Robotic devices have been intended to diminish the demands of manual-assistance repetitive tasks and further develop gait execution. scientific and clinical proof for the adequacy, safety, and tolerability of gait preparing with mechanical gadgets exists; notwithstanding, documentation of their benefits contrasted and customary treatments are restricted. This may be expected to some degree to the absence of proper patient selection and the parameters of locomotor training in light of functional disabilities.[37] The plan of the robot can't be precluded as a potential cause. Regardless of this shortcoming, mechanical gadgets are being incorporated into clinical settings with positive outcomes in a few applications.[38] Suitable use relies upon the clinician's information on various automated gadgets, just as the capacity to utilize the gadgets' technical elements, in this manner permitting patients to profit from robot-assisted gait preparing all through the restoration continuum with a definitive objective of returning to safe and effective overground walking.[39] There is expanding proof to help the idea of redesign and plasticity of the injured central nervous system (CNS). The potential for redesign is especially high early after CNS injury yet in addition conceivable at later stages.[40] Reorganization in a practically significant manner appears to rely upon the motor action as executed during rehabilitative preparation and followed by functional enhancements.[41,42] The science related to the practice of exercises in CNS issues is upheld by the therapy concept of an expanded dosage effect.[43] Task-oriented, high-repetition movements in light of the principles of motor learning can improve, in addition to other things, muscular strength, motor control, and movement coordination in patients with neurologic impairments.[44,45] Gait training can assist with forestalling secondary difficulties, for example, muscle atrophy, osteoporosis, bedsores, joint stiffness, and muscle-tissue shortening, and promotes the decrease of spasticity, among other benefits.[46] Robots upgrade the restoration process and may work on therapeutic results, also can possibly uphold clinical assessment by permitting instrumented measurement of physiologic and performance parameters, unequivocally control and measure the restorative intercession, execute novel types of mechanical manipulation impossible for the therapist to give, and supply various types of criticism, consequently expanding patient's motivation and further developing results.[47] Robots for rehabilitation were planned as a clinical instrument to computerize the work concentrated repetitive training strategies, particularly in the beginning phase of neurologic recovery where patients may need a high measure of assistance. Their programmable force-producing capacity permits robotic devices to advance task-oriented developments and give more right afferent input by directing the limb to advance heel strike and hip extension during initial contact and midstance of walking. Weight-supported treadmill training unweighs the patient and licenses the utilization of a mechanized treadmill for walking. Depending upon the patient's capacities and functional limit, up to 4 therapists may be needed to get and balance out a patient and guide the trunk and legs through a typical gait trajectory.[48] Robotic rehabilitation technology can expand the duration and number of training sessions while diminishing the quantity of helping therapists required. Robotic-aided gait training takes advantage of comparative elements of unweighing the lower limb and a mechanized treadmill, and substitutes for the physical work of the therapist with an exoskeleton automated framework that can reliably and dependably position and move the appendages during walking while giving a similar measure of degrees of freedom for the lower leg; knee; and, with ongoing specialized advances, the hip; and range of motion to advance a natural and comfortable gait. The robotic systems depicted here offer training conditions supportive of the standards for the enhancement of neuroplastic changes in patients with CNS-related gait impairment. The standards of intensity, repetition, task specificity and commitment are met in varying limits by these training gadgets. In the most recent 15 years, the number of robotic rehabilitation devices for upper and lower limbs has quickly expanded. For gait training, 4 significant robotic categories have been characterized:

* Tethered exoskeletons
* End-effector devices
* Untethered exoskeletons
* Patient-guided suspension systems

Tethered exoskeletal frameworks, like the Lokomat,[49] LOPES,[50] and ReoAmbulator,[51] apply powers through an inflexible articulated frame that move the patient's legs in at least 1 or more planes related to a bodyweight support framework. End-effector-based frameworks, like the Gait Trainer GT II[52] and G-EO frameworks (Reha Technologies, Switzerland),[53] work based on a limitation (i.e., power applied) at the distal finish of the kinetic chain that indicates the direction there and the proximal joints can essentially move as the body geometry and articulations dictate. Regularly, this implies the feet are strapped to 2 moving footplates, as in a curved trainer, and moved by the gadget in a gait like trajectory with less robotic control over the proximal joints.

Untethered exoskeletons, like the ReWalk Robotics, Indego, and the Ekso Bionics are wearable, fueled, articulated suits with independent power sources and control algorithms that permit the most freedom and realistic walking experience. Patient-directed suspension frameworks are non-anchored mechanical walking frames that permit overground, upheld walking through a moving frame with sensors and control calculations to help with propulsion and steering of the frame, and give a harness as well as trunk or pelvic support during walking. An illustration of these gadgets is the Andago, Hocoma AG, Switzerland. In spite of the fact that of incredible possible advantage and less expense.[54]

Tethered exoskeleton

Tethered exoskeletons have a device that encompasses the patient's legs, which might be suspended from an upward guide rail, upheld by a metal frame on wheels, or the exoskeleton can even be straightforwardly upheld by a mobile robot. stationary exoskeletons are normally associated straightforwardly to the ground through a rigid frame or directly bolted to the wall, improving and guaranteeing complete security. stationary and tethered exoskeletons can have enormous and strong engines and regulators, with none of their weight transferred to the client. These gadgets frequently include walking on a treadmill. Fastened exoskeletons will generally be to some degree less perplexing in their designing plan and more stable and innately more secure than gadgets that license overground walking attributable to the disposal of a fall hazard. A few gadgets have control frameworks that can detect and powerfully change execution assistance. One likely limit of these frameworks is that they are less obliging of individual gait varieties. Examples of tethered exoskeletons incorporate the Lokomat,[55] WalkTrainer,[56] NaTUre-gaits,[57] LOPES, ReoAmbulator, and the Anklebot. Some evidence focuses on the robotic part showing the same if not increased effectiveness when contrasted with more conventional physical therapy methods, while other evidence does not for instance in stroke individuals,[58,59] and affirms the long-standing thought that these gadgets really do decrease the physical requests on the therapist by giving partial bodyweight-upheld treadmill training. The biggest collection of scientific proof is for the Lokomat when utilized by people with a spinal cord injury (SCI) or stroke. This device is commercially accessible with a significant client base, permitting the consummation of clinical preliminaries. Regardless of this, or maybe in view of this, there is no agreement of whether and what it means for results in correlation with traditional or different sorts of robotic treatment.[60,61]

Motivation through a restorative gaming interface, conceivable in every one of the frameworks, expands patient commitment and resistance to treatment and decreases apparent distress, as shown in pediatric patients with cerebral palsy.[62] Labruyere and colleagues[63] measured game participation, utilizing electromyographic muscle action (Musculus rectus femoris) and heart rate during a demanding part and a less demanding part of the game. They concluded that kids with neurologic gait issues can change their activity to the demands incorporated into a virtual–reality experience situation. cognitive capacity and motor impedance decide how much they can do as such.

End-effector devices

End–effector gait trainers comprise of 2 footplates situated on 2 bars, 2 rockers, and 2 cranks, which give propulsive movement to the legs. The footplates create the stance and swing phases in many occurrences with symmetric movement. Ongoing improvements permit some level of deviation in the direction selected for every leg. The principle distinction contrasted with exoskeletons with a treadmill is that the feet are generally in touch with the moving platform, re-enacting the gait stages however not really creating genuine swing and stance stages. The directions of the footplates, as well as the vertical and horizontal movement of the center of mass, are programmable. The end–effector configuration fits gait retraining and step climbing. As a modality, this includes minimal interest for the client to have the option to start venturing movement furthermore may bring about more noteworthy changeability in the knee and hip movement, just as different levels of unweighing. Instances of end–effector gadgets incorporate G–EOGTII, and Lokohelp. A new little randomized, forthcoming review analysed 3 gait preparing strategies in people with traumatic brain injury. It inspected the effect of 3 unique methods of locomotor treatment on gait velocity and spatiotemporal balance utilizing an end–effector robot (G–EO), a robotic exoskeleton (Lokomat), and physically helped partial bodyweight-upheld treadmill training in people with traumatic brain injury Subjects went through 18 instructional sessions of 45 min long each. The training showed a measurably huge median expansion in self-chose velocity for all groups contrasted with pretraining. The most extreme velocity expanded for the Lokomat however not the G–EO bunch. The mobility part of the Stroke Impact Scale was fundamentally worked on just for the Lokomat bunch. The review examined the need of various staff furthermore high physical requests to give manual help for training. Staffing required for treatment arrangement was the least for the Lokomat.[64] In another comparative study, specialists took advantage of a clinical arrangement that put a Lokomat and a G-EO close to one another with a 3-layered kinematic recording framework. They acquired successive information from subjects with spinal cord or traumatic brain injury diagnosis utilizing the 2 frameworks and contrasted that and bodyweight-supported manually assisted treatment on a treadmill. The information affirmed a more controlled and tedious gait pattern when utilizing a Lokomat with a gait pattern that was generally like that of overground walking. The G-EO gave a step design that had a greater fluctuation of movement for the hips and knees, with marginally diminished knee movement, and the gait pattern varied somewhat from that seen during overground walking. At long last, the gait patterns accomplished during manually assisted treadmill bodyweight-upheld treatment were generally variable with the absence of symmetry of movement and timing.

Untethered exoskeletons

Untethered exoskeletons are bilateral robots that fuse actuators to move the patient's legs during the gait cycle, through a prearranged and near-normal gait cycle.[65] Their utilization normally requires the upper limb helps to keep up with the balance. They can likewise be partitioned into assistive and rehabilitative gadgets. The point of assistive devices is to work with mobility in the home and under community environmental conditions, though rehabilitative gadgets are planned to address the recuperation of gait function in patients with neurologic injuries. In devices like these, no less than 2 joints are actuated (hip and knee) and, typically, 1 joint (ankle) is precisely controlled. robotic hip-knee- ankle-foot exoskeletal orthoses have opened up and may help patients to stand; walk; and, sometimes, climb steps. These gadgets additionally have applications past mobility; for instance, exercise, improvement of secondary complications related to the absence of ambulation, and promotion of neuroplasticity.[66] Gait kinematics appear to differ broadly both across people within a device,[67] just as across devices. Studies have shown the chance of performing individual gait training in patients with an assortment of pathologic conditions, including SCI, traumatic brain injury, stroke, also, multiple sclerosis; the different CNS diagnoses for which exoskeletons are relevant were evaluated. As of now, for people with more noteworthy degrees of impairment, wheelchairs stay the favored mobility help yet still miss the mark looked at with upstanding bipedal walking. Untethered exoskeletons hold a lot of guarantees to satisfy this neglected need and have progressed considerably during the previous ten years as a suitable choice for both therapeutic and individual mobility purposes.

Patient-guided suspension systems

Patient-guided suspension frameworks are non-anchored overground rolling outlines that are intrinsically steady and provide a harness and additionally trunk or pelvic support, with dynamic unweighing for some, while permitting the client freedom to start and do the walking. They incorporate the Andago, SoloWalk,[68] KineAssist,[69] and WHERE-II.[70] Most have sensors and control algorithms to help with driving and moving of the frame and changing and additionally keeping up with stable support.[71] This sort of robot permits the patients to walk overground and to investigate the environment; patients are not bound to a proper region. In certain occurrences, the frame might be joined with untethered exoskeletons to offer the naturalistic autonomy of the latter with the safety of the former.

Upper limb rehabilitation

Around 795,000 strokes happen yearly in the United States, as per a report by the American Heart Association.[72] The worldwide situation of stroke is very like that in the United States; as per a report by the World Health Organization, consistently 15 million individuals experience a stroke around the world, among which only 65 percent survive.[73] The recovery in stroke events shows that a huge number of patients end up with disabilities; 25 percent of survivors are left with minor impairments, while 40 percent of survivors seem to encounter moderate-to-extreme disabilities requiring extraordinary care. Stroke is one of the primary reasons for serious long-term inability as it often impairs the upper limb functioning of the body. Apart from this, upper limb impairments can happen because of sports injuries, trauma, occupational injuries, and spinal cord injury. Since upper appendages are engaged with playing out a wide assortment of everyday tasks, it is basic to get autonomy once again to upper limb impaired patients as fast as could really be expected. Restoration programs are the primary technique for advancing utilitarian recuperation in people with upper limb dysfunction (ULD), requiring a long-term commitment by both the clinician and patient. To restore upper limb impairments, the broad task-oriented repetitive movement has ended up being a safe and viable technique that regularly relies upon one-on-one actual cooperation with the therapist as instances of ULD is expanding, robot-assisted therapy can possibly be a compelling arrangement in this respect. Besides, there are ongoing investigations substantiating that repetitive robot-assisted recovery programs decline upper limb motor impairment substantially.[74]

Robot-assisted treatment enjoys upper hands over conventional manual treatment, as the previous is equipped for giving treatment to patients to a more drawn-out timeframe is a more precise training technique and results in better quantitative criticism.[75] There are two sorts of robot-assisted rehabilitative gadgets in view of the mapping of a device's joint onto human anatomical joints (e.g., end-effector type and exoskeleton type). End-effector-type gadgets (e.g., MIT-MANUS-popularized as Inmotion,[76,77] Gentle/S,[78] ARM Guide)[79] are reasonable for end-point exercises as they can't give individual joint movement, meaning they can't map onto human anatomical joints.

Exoskeleton-type devices enjoy upper hands over end-effector-type devices, as they have unlimited authority over a patient's singular joint development and applied force, the better direction of movement, a moderately bigger range of motion (ROM), and better quantitative criticism. To date, various examination models of exoskeletons have been produced for human upper appendage restoration e.g., CADEN-7, SUFUL-7, ARMIN, RehabArm, BLUESABINO, and so on. Notwithstanding, their utilization in clinical settings at hospitals and outpatient centers is as yet restricted; just a single exoskeleton is industrially accessible for example ARMIN. To make an exoskeleton reasonable for clinical utilization, analysts have been searching for better arrangements as far as lightweight, minimization, low power-to-weight proportion, lightweight minimizers for power transmission, simple wear/doff, speedy wear/doff, alignment with human joints, modularity of kinematic structure, quick computation, sampling, control algorithm, and modularity of control. Notwithstanding rehabilitation, specialists have planned and created robotic exoskeletons for different applications too.

Existing upper limb exoskeleton

* ARMin-III[80,81] (replacement of ARMIn and ARMin-II, and ancestor of commercial exoskeleton Armeopower), created at ETH Zurich,

Switzerland, is one of the early and notable automated exoskeletons with high degrees of freedom for upper limb rehabilitation. The absolute first form ARMin[82] was planned with four DOFs expected to give recovery in the human shoulder (giving mobility for shoulder abduction/adduction, flexion–extension, and medial–lateral rotation) and elbow (flexion–extension). Then, at that point, the 7-DOF ARMin-II was created with five flexible lengths portions to give better tolerant cooperative recovery. Not at all like ARMin, the shoulder pivot of revolution isn't fixed in ARMin-II, permitting passive elevation/depression and protraction/retraction of the glenohumeral (GH) joint during shoulder vertical flexion–extension. ARMinII likewise incorporates ergonomic shoulder actuation to give much normal movement as could be expected for shoulder rehabilitation. The progression of the ARMin rehabilitative exoskeleton went through a few transformative phases and is currently monetarily accessible (known as ArmeoPower created by Hocoma AG, Volketswil, Switzerland) for use in human upper–limb recovery at clinical settings in hospitals.

- SUEFUL-7,[83] a 7-DOF upper limb motion-assist exoskeleton robot was created by Gopura et al. (2009) in Dr Kazuo Kiguchi's Laboratory at Saga University, Japan, in 2009. This robot was intended for giving movement assistance to genuinely frail people in their activities of everyday living. Design contemplations for SUEFUL-7 incorporate moving center of rotation (CR) of the shoulder joint and axes deviation of the wrist. In this robotic exoskeleton, the shoulder vertical and horizontal flexion/extension, elbow flexion/extension movements are driven by pulleys and cable drive while the actuators are put on the fixed edges. Though, for conveying shoulder internal/external rotation, forearm supination/pronation, wrist flexion/extension, and wrist radial/ulnar deviation movement to the robot. The actuators are connected to the robot itself and either straightforwardly associated or by means of gear drive. A slider-crank system was utilized for remunerating the CR of the shoulder joint. The robot weighs around 5 kg and is planned to be wheelchair mounted under the feeling that truly frail people use wheelchairs. To control SUEFUL-7, an EMG-based neuro-fuzzy control was applied. This exoskeleton was tried with healthy subjects only.

- ETS-MARSE,[84] a 7-DOF upper appendage exoskeleton for the entire arm, utilized a clever power transmission system to help shoulder internal/external rotation and forearm pronation supination.[85,86] Since it is to some degree hard to fit a shaft along the axis of rotation of the above cases (axis of humerus and radius), the engineer of ETS-MARSE utilized an antibacklash spur gear fit with open-type crescent gear and bearing assembly.

- Harmony,[87] a new mechanical exoskeleton in the field of upper limb recovery, has been created planning to empower the patient to do bilateral arm training. This framework is involved a double arm with a four-bar linkage, which makes it fit for giving natural shoulder movement. In contrast to, ARMin-III (where the shifting of shoulder CR was viewed as just for vertical flexion-extension), here the four-bar linkage component moved shoulder CR during either shoulder abduction/adduction or vertical flexion/extension, which made it more anatomical like[88,95,96]. The range of motion of the robot contrasts in view of the manner in which its different joints are designed. For example, ROM of shoulder abduction increased when it was performed all the while with shoulder external rotation.

- CABXLexo-7,[89] a 7-DOF cable-driven upper-limb exoskeleton (CABXLexo-7) replacement to 6-DOF CABexo,[90] was created by Feiyun et al. To make a lightweight exoskeleton robot fit for giving each of the seven DOFs, the development group set every one of the actuators on the fixed board and sent power through a cable-conduit framework utilizing two kinds of cable-driven differential components and utilizing a pressure gadget to work with the cable slag issue. The subsequent load of the moving robot is 3.5 kg. Be that as it may, this robot doesn't think about the movement of CR of the human shoulder. Trial and error were completed utilizing surface electromyography with five sound people. This robot was intended for giving movement help to post-stroke patients. CABXLexo-7 is yet to go through clinical preliminaries.

Discussion

Throughout the last two decades, various exoskeletons have been created to restore individuals with upper and lower limb disabilities and scientists have been broadly and continually attempting to propel the hardware design and control approaches for such robot-assisted therapeutic devices. The speed of advancement in these robotic advances keeps on expanding, as well as to specialize. Robotic technologies keep on being assessed and show basically unassuming advantages across an assortment of pathologic conditions; for instance, kids with cerebral palsy,[91] stroke, SCI,[92] Parkinson's disease, Brown-Sequard syndrome, and vascular dementia. It stays hazy, nonetheless, regardless of whether robot-helped treatment is better in specific clinical situations than conventional treatment. At last, apparently, the main outcome measure, Quality of life, has started to be studied as well. However, as a rule, there have been many reports of solid positive results utilizing Robotics, the criteria as well as the methodology for getting ideal advantages from robotics in the rehabilitation the field is still a long way from well-described[93]; in any case, there is likewise a developing work to bring lucidity to this field.[94] In spite of the way that a tremendous measure of exploration has been done, the improvement of a control technique to give successful restoration is still developing. To give viable restoration to upper and lower limb impaired patients, first, we should decide the patient's protected outrageous ROM previously before starting the treatment. This may likewise assist with choosing proper and safe restoration conventions for patients. For instance, it is fundamental to know the safe ROM of a patient having muscle tone. In conventional treatment, this is done physically by a therapist who directly observes the patient. Second, the therapist can change his/her methodology in the event that the patient feels any aggravation or something undesirable occurs. This is the kind of thing exoskeletons ought to remember for their control design.

Conclusion

In this section, the hardware design, safety, compactness, control method, actuation, and power transmission mechanism of existing upper/lower limb rehabilitative exoskeletons were inspected. We tracked down that most research protypes of existing exoskeletons were not moved into a commercial item, and maybe the explanation for it is the absence of clinical testing. Accordingly, in this review, the challenges that should be tended to for further developed functionality has been distinguished and examined. Addressing these difficulties in the advancement of exoskeletons will make them more practical for rehabilitation purposes. Moreover, recommendations were given as appropriate to work on the functionality of exoskeletons.

References

1. Obeso JA, Rodríguez-Oroz MC, Benitez-Temino B, Blesa FJ, Guridi J, Marin C. Functional organization of the basal ganglia: therapeutic implications for Parkinson's disease. *Rodriguez MMov Disord*. 2008;23(Suppl 3):S548–S559.
2. Voitenkov V, Smirnov N, ˇkusheva E, Skripchenko N. Neurophysiology methods in the assessment of the efficacy of rehabilitation of sensomotor disturbances due to spinal cord lesions. *J Neurol Sci*. 2017;381:1098. doi:10.1016/j.jns.2017.08.3100.

3. Hatem SM, Saussez G, Della Faille M, et al. Rehabilitation of Motor Function after Stroke: a Multiple Systematic Review Focused on Techniques to Stimulate Upper Extremity Recovery. *Front Hum Neurosci.* 2016;10:442.

4. Bello-Haas VD. Physical therapy for individuals with amyotrophic lateral sclerosis: current insights. *Degener Neurol Neuromuscul Dis.* 2018;8:45–54.

5. Veerbeek JM, van Wegen E, van Peppen R, et al. What is the evidence for physical therapy poststroke? A systematic review and meta-analysis. *PLoS One.* 2014;9(2):e87987.

6. Gassert R, Dietz VJ. Rehabilitation robots for the treatment of sensorimotor deficits: a neurophysiological perspective. *Neuroeng Rehabil.* 2018 Jun 5;15(1):46.

7. Turner DL, Ramos-Murguialday A, Birbaumer N, Hoffmann U, Luft A. Neurophysiology of robot-mediated training and therapy: a perspective for future use in clinical populations. *Front Neurol.* 2013 Nov 13;4:184.

8. Calabrò RS, Portaro S, Manuli A, Leo A, Naro A. Rethinking the robotic rehabilitation pathway for people with amyotrophic lateral sclerosis: a need for clinical trials. *Bramanti AInnov Clin Neurosci.* 2019 Jan 1;16(1–2):11–12.

9. Paganoni S, Karam C, Joyce N, Bedlack R, Carter GT. Comprehensive rehabilitative care across the spectrum of amyotrophic lateral sclerosis. *NeuroRehabilitation.* 2015;37(1):53–68.

10. Ferrazzoli D, Ortelli P, Zivi I, et al. Efficacy of intensive multidisciplinary rehabilitation in Parkinson's disease: a randomised controlled study. *Neurol Neurosurg Psychiatry.* 2018 Aug;89(8):828–835.

11. Tomlinson CL, Patel S, Meek C, et al. Physiotherapy versus placebo or no intervention in Parkinson's disease. *Cochrane Database Syst Rev.* 2012 Jul 11(7).

12. Prodoehl J, Rafferty MR, David FJ, et al. Two-year exercise program improves physical function in Parkinson's disease: the PRET-PD randomized clinical trial. *Neurorehabil Neural Repair.* 2015 Feb;29(2):112–122.

13. Shirota C, van Asseldonk E, Matjačić Z, et al. Robot-supported assessment of balance in standing and walking. *J Neuroeng Rehabil.* 2017 Aug 14;14(1):80.

14. Krebs HI, Volpe BT. Rehabilitation robotics. *Handb Clin Neurol.* 2013;110:283–294.

15. Morris ME, Martin CL, Schenkman ML. Striding out with Parkinson disease: evidence-based physical therapy for gait disorders. *Phys Ther.* 2010 Feb;90(2):280–288.

16. Major K, Major Z, Carbone G, et al. Ranges of Motion as Basis for Robot-Assisted Post-Stroke Rehabilitation. Hum. *Vet Med.* 2016;8:192–196.

17. Vaida C, Carbone G, Maior K, Maior Z, Plitea N, Pisla D. On Human Robot Interaction Modalities in the Upper Limb Rehabilitation after Stroke. *ACTA Tech Napoc Ser Appl Math Mech Eng.* 2017;60:91–102.

18. A. H. Association. *Heart Disease and Stroke Statistics −2005 Update.* Dallas, Texas: American Heart Association; 2005.

19. Hogan N, Krebs H. Interactive robots for neuro-rehabilitation. *Restor Neurol Neurosci.* 2004;22:349–358.

20. Tenzer Y, Jentoft LP, Howe RD. The Feel of MEMS Barometers: inexpensive and Easily Customized Tactile Array Sensors. *IEEE Robotics & Automation Magazine.* 2014;9:89–95.

21. Shimojo M, Namiki A, Ishikawa M, Makino R, Mabuchi K. A tactile Sensor sheet using pressure conductive rubber with electrical-wires Stitched method. *IEEE Sens J.* 2004;4(5):589–596.

22. Aqilah A, Jaffar A, Bahari S, Low CY, Koch T. Resistivity characteristics of single miniature tactile sensing element based on pressure Sensitive conductive rubber sheet. *Proceedings of the IEEE 8th International Colloquium on Signal Processing and Its Applications (CSPA'12);* March 2012:223–227.

23. Battaglia E, Bianchi M, Altobelli A, et al. ThimbleSense: a Fin-gertip-Wearable Tac-tile Sensor for Grasp Analysis. *IEEE Trans Haptics.* 2016;9(1):121–133.

24. Tsai TH, Tsai HC, Wu TK. A CMOS micromachined capacitive tactile sensor with integrated readout circuits and compensation of process Variations. *IEEE Transaction on Biomedical Circuits and Systems.* 2014;8(5):608–616.

25. Oess NP, Wanek J, Curt A. Design and evaluation of a low-cost instrumented glove for hand function assessment. *J Neuroeng Rehabil*. 2012;9:2–11.
26. Dipietro L, Sabatini AM, Dario P. A survey of glove-based systems And their applications. *IEEE Transaction on System Man and Cybernatics*. 2008;38(4):461–482.
27. Brage S, Westgate K, Franks PW, Stegle O, Wright A, Ekelund U, et al. Estimation of free-living energy expenditure by heart rate and movement sensing: a doubly-labelled water study. *PLoS One*. 2015;10:1–20. doi:10.1371/journal.pone.0137206.
28. Altini M, Penders J, Vullers R, Amft O. Estimating energy expenditure using Body-worn accelerometers: a comparison of methods, sensors number and positioning. *IEEE J Biomed Heal Inform*. 2015;19:219–226. doi:10.1109/JBHI.2014.2313039.
29. Williams KR. The relationship between mechanical and physiological estimates. *Med Sci Sports Exerc*. 1985;17:317–325.
30. Cavagna GA, Kaneko M. Mechanical work and efficiency in level walking and Running. *J Physiol*. 1977;268:467–481. doi:10.1113/jphysiol.1977.sp011866.
31. Chernbumroong S, Cang S, Yu H. A practical multi-sensor activity recognition System for home-based care. *Decis Support Syst*. 2014;66:61–70. doi:10.1016/j.Dss.2014.06.005.
32. Brage S, Westgate K, Franks PW, Stegle O, Wright A, Ekelund U, et al. Estimation of free-living energy expenditure by heart rate and movement sensing: a Doubly-labelled water study. *PLoS One*. 2015;10:1–20. doi:10.1371/journal.pone.0137206.
33. Nathan D, Huynh DQ, Rubenson J, Rosenberg M. Estimating physical activity energy expenditure with the kinect sensor in an exergaming environment. *PLoS One*. 2015;10:1–22. doi:10.1371/journal.pone.0127113.
34. Calabro RS, Cacciola A, Berte F, et al. Robotic gait rehabilitation and substitution devices in neurological disorders: where are we now? *Neurol Sci*. 2016;37(4):503–514.
35. Reinkensmeyer DJ, Emken JL, Cramer SC. Robotics, motor learning, and neurologic recovery. *Annu Rev Biomed Eng*. 2004;6:497–525.
36. Esquenazi A, Lee S, Packel AT, et al. A randomized comparative study of manually assisted versus robotic-assisted body weight supported treadmill training in persons with a traumatic brain injury. *PM R*. 2013;5(4):280–290.
37. Krakauer JW, Carmichael ST, Corbett D, et al. Getting neurorehabilitation right: what can be learned from animal models? *Neurorehabil Neural Repair*. 2012;26(8):923–931.
38. Kleim JA, Jones TA. Principles of experience-dependent neural plasticity: implications for rehabilitation after brain damage. *J Speech Lang Hear Res*. 2008;51(1):S225–S239.
39. Murphy TH, Corbett D. Plasticity during stroke recovery: from synapse to behaviour. *Nat Rev Neurosci*. 2009;10(12):861–872.
40. Dietz V. Neuronal plasticity after a human spinal cord injury: positive and negative effects. *Exp Neurol*. 2012;235(1):110–115.
41. Edgerton VR, Tillakaratne NJ, Bigbee AJ, et al. Plasticity of the spinal neural circuitry after injury. *Annu Rev Neurosci*. 2004;27:145–167.
42. Maier IC, Schwab ME. Sprouting, regeneration and circuit formation in the injured spinal cord: factors and activity. *Philos Trans R Soc Lond B Biol Sci*. 2006;361(1473):1611–1634.
43. Dromerick AW, Lum PS, Hidler J. Activity-based therapies. *NeuroRx*. 2006;3(4):428–438.
44. Dietz V, Harkema SJ. Locomotor activity in spinal cord-injured persons. *J Appl Physiol. (1985)*. 2004;96(5):1954–1960.
45. Kwakkel G, Wagenaar RC, Twisk JW, et al. Intensity of leg and arm training after primary middle-cerebral-artery stroke: a randomised trial. *Lancet*. 1999;354(9174):191–196.
46. Edgerton VR, de Leon RD, Tillakaratne N, et al. Use-dependent plasticity in spinal stepping and standing. *Adv Neurol*. 1997;72:233–247.
47. Esquenazi A, Packel A. Robotic-assisted gait training and restoration. *Am J Phys Med Rehabil*. 2012;91(11 Suppl 3):S217–S227 [quiz: S228–31].
48. Iosa M, Morone G, Fusco A, et al. Seven capital devices for the future of stroke rehabilitation. *Stroke Res Treat*. 2012;2012.

49. Colombo G, Joerg M, Schreier R, et al. Treadmill training of paraplegic patients using a robotic orthosis. *J Rehabil Res Dev.* 2000;37(6):693–700.
50. Veneman JF, Kruidhof R, Hekman EE, et al. Design and evaluation of the LOPES exoskeleton robot for interactive gait rehabilitation. *IEEE Trans Neural Syst Rehabil Eng.* 2007;15(3):379–386.
51. Mantone J. Getting a leg up? Rehab patients get an assist from devices such as HealthSouth's AutoAmbulator, but the robots' clinical benefits are still in doubt. *Mod Healthc.* 2006;36(7):58–60.
52. Hesse S, Uhlenbrock D, Werner C, et al. A mechanized gait trainer for restoring gait in nonambulatory subjects. *Arch Phys Med Rehabil.* 2000;81(9):1158–1161.
53. Hesse S, Waldner A, Tomelleri C. Innovative gait robot for the repetitive practice of floor walking and stair climbing up and down in stroke patients. *J Neuroeng Rehabil.* 2010;7:30.
54. Esquenazi A, Maier IC, Schuler TA, et al. Clinical application of robotics and technology in the restoration of walking. In: D R, V D, eds. *Neurorehab Technol.* Springer; 2016:223–248.
55. Colombo G. The "Lokomat"-A driven ambulatory orthosis. *Germany: dizinich Orthopadesche Technik.* 2000;6:178–181.
56. Allemand Y, Stauffer Y, Clavel R, et al. Design of a new lower extremity orthosis foroverground gait training with the WalkTrainer. Paper presented at: 2009 IEEE International Conference on Rehabilitation Robotics. Kyoto, Japan, June 23–26, 2009.
57. Wang P, Low KH, Tow A, et al. Initial system evaluation of an overground rehabilitation gait training robot (NaTUre-gaits). *Adv Robot.* 2011;25(15):1927–1948.
58. Nam YG, Lee JW, Park JW, et al. Effects of electromechanical exoskeletonassisted gait training on walking ability of stroke patients: a randomized controlled trial. *Arch Phys Med Rehabil.* 2019;100(1):26–31.
59. Hornby TG, Campbell DD, Kahn JH, et al. Enhanced gait-related improvements after therapist-versus robotic-assisted locomotor training in subjects with chronic stroke: a randomized controlled study. *Stroke.* 2008;39:1786–1792.
60. Mayr A, Kofler M, Quirbach E, et al. Prospective, blinded, randomized crossover study of gait rehabilitation in stroke patients using the Lokomat gait orthosis. *Neurorehabil Neural Repair.* 2007;21(4):307–314.
61. Hidler J, Nichols D, Pelliccio M, et al. Multicenter randomized clinical trial evaluating the effectiveness of the Lokomat in subacute stroke. *Neurorehabil Neural Repair.* 2009;23(1):5–13.
62. Brutsch K, Koenig A, Zimmerli L, et al. Virtual reality for enhancement of robotassisted gait training in children with central gait disorders. *J Rehabil Med.* 2011;43(6):493–499.
63. Labruyere R, Gerber CN, Birrer-Brutsch K, et al. Requirements for and impact of a serious game for neuro-pediatric robot-assisted gait training. *Res Dev Disabil.* 2013;34(11):3906–3915.
64. Esquenazi A, Lee S, Wikoff A, et al. A comparison of locomotor therapy interventions: partial-body weight-supported treadmill, Lokomat, and G-EO training in people with traumatic brain injury. *PM R.* 2017;9(9):839–846.
65. Molteni F, Gasperini G, Gaffuri M, et al. Wearable robotic exoskeleton for overground gait training in sub-acute and chronic hemiparetic stroke patients: preliminary results. *Eur J Phys Rehabil Med.* 2017;53(5):676–684.
66. Esquenazi A, Talaty M, Jayaraman A. Powered exoskeletons for walking assistance in persons with central nervous system injuries: a narrative review. *PM R.* 2017;9(1):46–62.
67. Talaty M, Esquenazi A, Briceno JE. Differentiating ability in users of the ReWalk(TM) powered exoskeleton: an analysis of walking kinematics. *IEEE Int Conf Rehabil Robot.* 2013;2013.
68. McCormick A, Alazem H, Morbi A, et al. Power Walker helps a child with cerebralpalsy. 3rd International Conference on Control, Dynamic Systems, and Robotics. Ottawa, Canada, May 9–10, 2016.

69. Patton J, Brown DA, Peshkin M, et al. KineAssist: design and development of a robotic overground gait and balance therapy device. *Top Stroke Rehabil.* 2008;15(2):131–139.
70. Seo KH, Lee JJ. The development of two mobile gait rehabilitation systems. *IEEE Trans Neural Syst Rehabil Eng.* 2009;17(2):156–166.
71. Alias NA, Huq MS, Ibrahim BSKK, et al. The efficacy of state of the art overground gait rehabilitation robotics: a bird's eye view. *Procedia Comput Sci.* 2017;105:365–370.
72. Accogli A, Grazi L, Crea S, et al. EMG-based detection of user's intentions for human-machine shared control of an assistive upper-limb exoskeleton. In: González-Vargas J, Ibáñez J, Contreras-Vidal JL, van der Kooij H, Pons J, eds. *Wearable Robotics: Challenges and Trends.* Cham: Springer International Publishing; 2017:181–185.
73. American Stroke Association, 2019. https://www.stroke.org/we-can-help/survivors/stroke-recovery/first-steps-to-recovery/rehabilitation-therapy-after-a-stroke/.
74. Amirabdollahian F, Loureiro R, Gradwell E, Collin C, Harwin W, Johnson G. Multivariate analysis of the Fugl-Meyer outcome measures assessing the effectiveness of GENTLE/S robot-mediated stroke therapy. *J Neuroeng Rehabil.* 2007;4(1):4.
75. Teasell RW, Kalra L. What's new in stroke rehabilitation. *Stroke.* 2004;35(2):383–385. https://doi.org/10.1161/01.str.0000115937.94104.76.
76. Hogan N, Krebs HI, Charnnarong J, Srikrishna P, Sharon A. MIT-MANUS: a workstation for manual therapy and training. I. *Proceedings IEEE International Workshop on Robot and Human Communication;* 1992:161–165.
77. Krebs HI, Volpe BT, Williams D, et al. Robot-aided neurorehabilitation: a robot for wrist rehabilitation. *IEEE Trans Neural Syst Rehabil Eng.* 2007;15(3):327–335. https://doi.org/10.1109/TNSRE.2007.903899.
78. Coote S, Murphy B, Harwin W, Stokes E. The effect of the GENTLE/s robotmediated therapy system on arm function after stroke. *Clin Rehabil.* 2008;22(5):395.
79. Reinkensmeyer DJ, Kahn LE, Averbuch M, McKenna-Cole AN, Schmit B, Rymer W. Understanding and treating arm movement impairment after chronic brain injury: progress with the ARM guide. *J Rehabil Res Dev.* 2000;37(6):653–662.
80. a Nef T, Guidali M, Klamroth-Marganska V, Riener R. Armin—Exoskeleton robot for stroke rehabilitation. In: Dossel O, Schlegel WC, eds. *World Congress € on Medical Physics and Biomedical Engineering, September 7–12, 2009.* Munich, Germany: Springer, Berlin, Heidelberg; 2009:127–130.
81. Nef T, Guidali M, Riener R. ARMin III—Arm therapy exoskeleton with an ergonomic shoulder actuation. *Appl Bionics Biomech.* 2009;6(2).
82. Nef T, Mihelj M, Kiefer G, Perndl C, Muller R, Riener R. Arminexoskeleton for arm therapy in stroke patients. *IEEE 10th International Conference on Rehabilitation Robotics, 2007. ICORR 2007* IEEE; 2007:68–74.
83. Gunasekara M, Gopura R, Jayawardena S. 6-REXOS: upper limb exoskeleton robot with improved pHRI. *Int J Adv Robot Syst.* 2015;12(4):47. https://doi.org/10.5772/60440.
84. Rahman MH, Saad M, Kenne J- P, Archambault PS. Control of an exoskeleton robot arm with sliding mode exponential reaching law. *Int J Control Autom Syst.* 2013;11(1):92–104.
85. Rahman MH, Kittel-Ouimet T, Saad M, Kenne J- P, Archambault PS. Development and control of a robotic exoskeleton for shoulder, elbow and forearm movement assistance. *Appl Bionics Biomech.* 2012;9(3).
86. Rahman MH, Rahman MJ, Cristobal OL, Saad M, Kenne JP, Archambault PS. Development of a whole arm wearable robotic exoskeleton for rehabilitation and to assist upper limb movements. *Robotica.* 2014;33(1):19–39.
87. Kim B, Deshpande AD. Controls for the shoulder mechanism of an upper-body exoskeleton for promoting scapulohumeral rhythm. *2015 IEEE International Conference on Rehabilitation Robotics (ICORR);* 2015:538–542.
88. Gupta M, Rajak BL, Bhatia D, Mukherhjee A. Neuromodulatory effect of repetitive transcranial magnetic stimulation pulses on functional motor performances of spastic cerebral palsy children. *J Med Eng Technol.* 2018.

89. Xiao F, Gao Y, Wang Y, Zhu Y, Zhao J. Design and evaluation of a 7-DOF cable-driven upper limb exoskeleton. *J Mech Sci Technol.* 2018;32(2):855–864. https://doi.org/10.1007/s12206-018-0136-y.

90. Xiao F, Gao Y, Wang Y, Zhu Y, Zhao J. Design of a wearable cable-driven upper limb exoskeleton based on epicyclic gear trains structure. *Technol Health Care.* 2017;25(S1):3–11. https://doi.org/10.3233/THC-171300.

91. Carvalho I, Pinto SM, Chagas DDV, et al. Robotic gait training for individuals with cerebral palsy: a systematic review and meta-analysis. *Arch Phys Med Rehabil.* 2017;98(11):2332–2344.

92. Esquenazi A, Talaty M, Packel A, et al. The ReWalk powered exoskeleton to restore ambulatory function to individuals with thoracic-level motor-complete spinal cord injury. *Am J Phys Med Rehabil.* 2012;91(11):911–921.

93. van Hedel HJA, Severini G, Scarton A, et al. Advanced Robotic Therapy Integrated Centers (ARTIC): an international collaboration facilitating the application of rehabilitation technologies. *J Neuroeng Rehabil.* 2018;15(1):30.

94. Bayon C, Martin-Lorenzo T, Moral-Saiz B, et al. A robot-based gait training therapy for pediatric population with cerebral palsy: goal setting, proposal and preliminary clinical implementation. *J Neuroeng Rehabil.* 2018;15(1):69.

95. Kim B, Deshpande AD. An upper-body rehabilitation exoskeleton Harmony with an anatomical shoulder mechanism: Design, modeling, control, and performance evaluation. *Int J Robot Res.* 2017;36. doi:10.1177/0278364917706743.

96. Park BJ, Hunt SJ, Nadolski GJ, et al. 3D Augmented reality-assisted CT-Guided interventions: system design and preclinical trial on an abdominal phantom using HoloLens 2, 2020. arXiv preprint arXiv:2005.09146 (2020).

CHAPTER 6

Neural prosthesis in rehabilitation

Introduction to the mechanism of action of neural prosthesis in rehabilitation

The complex and one of the most salient organs of the human body is the brain. It is responsible for senses, movement, and control, emotions and feelings, language and communication, thinking and memory. The brain is protected by the bony skull and it is located at the anterior end. The brain determines how one response to stressful situations by regulating the heart and breathing rate. The brain is connected to the spinal cord in the lower part. The adult human brain weighs between 1,200 to 1,500 gm and contains about one trillion cells called neurons that are connected by trillions of connections, or synapses. It occupies a volume of about 1400 cc- approximately 2 percent of the total body weight and receives 20 percent of the blood, oxygen, and calories supplied to the body. The brain along with the spinal cord comprises the central nervous system (CNS) and it is the primary command. The brain along with the spinal cord comprises the central nervous system (CNS) and is the primary command center of the body. The spinal cord is a bundle of nerve fibers running in two pathways between the neck and lower back under the well-protecting spinal column. The adult spinal cord is approximately 40 to 50 cm long and occupies about 150 cc.[1,2] The spinal cord nerves transmit information from body organs and external stimuli to the brain and send information from the brain to other areas of the body. The nerve bundles travel in two pathways- ascending nerve tracts carry sensory information from the body to the brain and descending nerve tracts send information about the motor function from the brain to the rest of the body.[3] The CNS in its function is assisted by the peripheral nervous system (PNS) which is comprised of the autonomic nervous system and the somatic nervous system. The autonomic has involuntary control of internal organs, blood vessels, and smooth and cardiac muscles which controls vital functions such as breathing, digestion, heart rate, and secretions of hormones. The somatic has voluntary control of skin, bone joints, and skeletal muscle. The two systems functions together by the way of nerves from the PNS entering and becoming part of the CNS, and vice versa.

Modern Intervention Tools for Rehabilitation.
DOI: https://doi.org/10.1016/B978-0-323-99124-7.00006-7

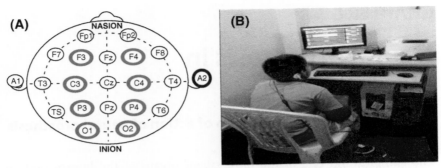

Figure 6.1 (A) Montage 10–20 showing EEG recording areas; (B) EEG recording of CP child.

The interplay between the CNS and PNS allows an individual to sense the environmental stimuli and integrate the information respond appropriately and carry out a different range of behaviors and this ultimately accounts for one of the major systems of the human body the nervous system which, to be simplified, is a network of specialized cells called neurons that transmit information in the form of electrical signals between different parts of the body to coordinate its actions.[4] According to many studies, chronic stress damages brain function in multiple ways. It can disturb the synapse regulation, resulting in the loss of sociability and the avoidance of interactions with others. Stress can deplete the brain cells and reduce brain size. Chronic stress has a shrinking effect on the prefrontal cortex, the area of the brain responsible for memory and learning. Therefore, stress plays a pivotal role in the remodeling of the brain structure as shown in Fig. 6.1.

Neural prosthetics combine neuroscience with biomedical engineering to develop devices to provide lost functionality. Sometimes they are confused with the brain–computer interface (BCI) which connects the brain to a computer. Neural prosthetics are a combination of neurophysiology with electrophysiology as shown in Fig. 6.7. These assistive devices play an important role to restore functions lost as a result of neural damage. They can be externally or internally implanted and electrically stimulate the nerves and restore lost functionality. The spinal cord and deep brain stimulators are routinely implanted to restore limb movements with help of electrodes under the skin. The neural prosthetics can substitute for a motor, sensory or cognitive impairment due to an injury or disease. They can be used to enhance functionality in damaged organs by providing a closed-loop system wherein signals from the brain could be detected to cause desired motor response and feedback from sensory organs can be sent to the brain for

| Sensory and Motor Signals from Brain | Data Acquisition and Hardware control system | Feature Extraction Algorithm | Output to Neural prosthetic Limb |

Figure 6.2 Simplified Neural Prosthetics System.

encoding as shown in Fig. 6.2. A common example of neural prosthetics is a cochlear implant which is employed to restore auditory loss in individuals. Through the replacement or augmentation of damaged senses, these devices intend to improve the quality of life for those with disabilities.[5]

The neural prosthetics are designed to be employed in the animal research studies wherein these implanted devices monitor the daily activities and transmit information over a wireless network to the connected system for continuous monitoring. It allows for studying neuron-to-neuron interaction at a local level. These minimally invasive devices are designed to assist the lost functionality of regions surrounding the brain, eyes, and cochlea and communicate with their surroundings wirelessly. To avoid damage to the surrounding tissue due to inflammation or burns, the power consumption requirements for such devices should be minimal.[6,7]

Cognitive, visual, auditory, and prosthetic hands are some examples of neural prostheses which are employed to restore lost functionality in the brain, eye, ear, and limb extremities respectively. The field has seen widespread growth in the domain of wearable sensors which are driving innovation. These devices can be employed to monitor the health condition of chronic patients and the elderly in remote or at-home settings. Wearable sensors have been employed in diagnostic, physiological, motion, and biochemical sensing. This could help in the diagnosis and treatment of persons suffering from cardiovascular, neurological, and pulmonary diseases. Further, they can help in improving the independence and quality of life of elderly persons and those suffering from disabilities by preventing falls, and improved gait, and mobility. Although issues may arise due to connectivity and certain local factors such as remoteness of a location, technological barriers and stigma associated with the use of the technology could limit its potential. However, wearable technology has the potential to allow the reach specialists from urban areas to remote rural areas and enhance the quality of healthcare services.[6,8]

One of the leading cause of disability is Traumatic Brain Injury (TBI) which causes huge burden on the individual, their family members and the society at large. The patients with mild TBI have their movement restricted and live with limited functions post disability for several years. Despite high incidents of brain injury, the treatment options available are limited which may be a barrier to provide effective treatment to the survivors. With enhanced cognitive performance in the TBI patients could allow them to return to their daily functioning and restore the quality of life of such individuals. However, the relationship between the cognitive impairment and everyday routine functioning is not fully understood yet and requires advanced research. The role of cognition in higher levels of functional recovery is still unclear and is being studied. It is believed that cognitive impairment is a key driver of disability, but gaps exists in understanding the relationship between cognitive impairment and cognition. Hence, it is important to understand the important cognitive domains related to daily human functioning and establish the cognitive impairments that impact the functioning of human brain at different levels. Further, the overall emotional and mental health factors play a vital role in functional recovery of patients affected by TBI. With the help of neural prosthetics, devices could be developed to reduce cognitive impairment and enhance cognition in people affected with TBI.[7–9]

Clinical application relevance in sensory prosthesis

The deployment of technological interventions could help in overcoming neurological deficits by incorporating principles from multi-disciplinary fields such as electrical, biomedical, mechanical, and materials engineering and advanced fields such as neuroscience, polymer science, and electro-chemistry. The field underlies the principle of artificial manipulation of the biological neural system by externally inducing electric fields and mimicking normal sensorimotor functions. However, the challenges lie in the development of appropriate engineering hardware which can be implanted to carry out the desired task or function in the human body. The electrical stimulation applied tends to replicate the normal temporal-spatial firing patterns of neurons. However, achieving the correct firing pattern of neurons is a challenging task due to the limitation of present technology and the availability of sophisticated miniaturized electrodes for targeted stimulation. The neural prosthesis may be classified into sensory and motor. The motor prosthesis employs electrical stimulation to the neuro–muscular system to provide an alternative for normal control of my brain or the spinal cord which is affected due to an injury or disease. Whereas sensory prosthetics employ the use of artificial sensors to replace neural input that comes from a peripheral biological source. With the advancement in technology, these

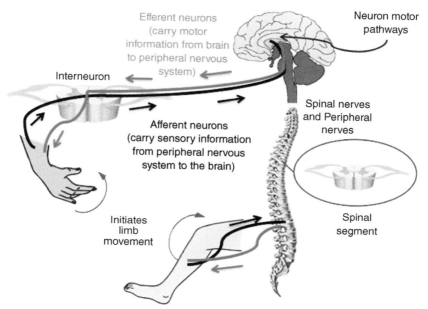

Figure 6.3 Physiology of motor control pathway.[29]

neural interfaces would directly control the implanted devices to take over the function of damaged body organs such as the brain or the spinal cord as shown in Fig. 6.3.[9,10]

The advanced auditory and visual prosthetics are bringing a sense of hope and joy to the lives of the disabled especially the younger population and children to perform their tasks of daily living with much comfort and ease. These devices artificially stimulate neural tissue to replicate absent biological stimuli. For both auditory and visual prosthesis, several challenges with regards to the development of implantable devices and external hardware need to be addressed along with the requisite technological tools for the stimulation of neural system, signal processing, electronic design fabrication techniques, ethical approval for smoother clinical implementation of these devices in humans.[7,10]

Sensory feedback in neuro-prosthetics is a difficult task to achieve but allows to rapidly transform the life of an amputee and disabled population. From conventional replacement of limbs to modern advanced sensory prosthetics, the technology has significantly changed in the last 20 years. With better degrees of freedom and range these sophisticated and smart brain-controlled devices allow decoding of the brain massage and control

Figure 6.4 Cause of loss of limb control.

of limb function to allow range of body movements which were previously not possible. The field of research has become quite dynamic and vibrant in the modern era which has caught the attention and raised curiosity in their minds to dwell into the new field. The electrodes in the motor cortex region of the brain of the paralysed individual allow us to understand the firing pattern of the muscles and associated limb movement. It allows better inference of patient's movement and designing of prosthetic arm. However, the neuro–prosthetics suffer from allowing the paralysed persons the feeling or sense of touch and surface texture which is usually taken for granted. Without the sense of touch, even though the motor and sensory unit is intact, the patient cannot distinguish between objects and this makes the whole exercise futile and non–beneficial for the subject. Several institutes and industries are spending huge amount of money in carrying out further research the area of haptics and touch sensation to improve the present designs and make them look more natural and appealing to the patients as shown in Fig. 6.4. They are actively studying the functioning of existing nervous system and trying to exactly reproduce the same by electrically stimulating different body structures in real-time mode. With help of inputs and responses from the disabled population, the researchers are deciphering the language of human touch and how they can be integrated in the neural prosthetics designs.[7,8,11] The development of prosthetic hand research is driven by clinical needs to achieve desired functionality and enhanced outcome measures. Further, restoring a sense of touch can help the amputees to improve the use of attached prosthetic hands with improved control and sensory feedback. The appearance and device control are important features which determine the user preference for the device and its long term reliability.

Researchers are employing new techniques such as Targeted Muscle Reinnervation (TMR) and Osseointegration to develop enhanced signal decoding strategies. Further, designing of control strategies that enable simultaneous and proportional control of prosthesis which allows better dexterity and user control of the prosthetic limb. Moreover, with advent of advanced machine learning (ML) and artificial intelligence (AI) techniques would allow to resolve the issues related to device command and control. Sensory feedback is also important for the control of prosthetic hand. The amputees usually rely on visual and auditory information to monitor their prosthesis during manipulation. With recent advances in providing sensory feedback, researchers provide complete feedback loop by providing natural sensations thereby improving user experience and helping in better device experience.[6,7,10] With advanced materials, sophisticated electronic skins (e-skins) are being developed which provide more flexibility with ability to have touch perception and proprioception. The e-skin are self-healing, stretchable, can measure pressure and temperature by showing natural human skin-like characteristics.

By understanding the firing patterns of different motor fibres, researchers are unravelling the mysterious functioning of the human brain and touch perception, however it is still far to achieve the perfect distinguish between the egg and cloth. New research could unravel such mystery that may allow better touch sensitive neural devices in the near future. With advancement in wireless technology and availability of bionic prosthetics, new devices would allow better biocompatibility and flexibility of implanted electrode arrays. With collaboration of industry and academia it would be possible to develop such sophisticated devices in the near future which may possibly work better than existing devices and improve the quality of life of persons

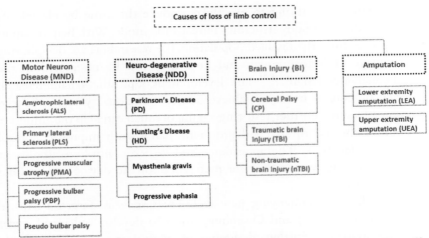

Figure 6.5 Summarized representation of various contributors in limb loss control.[29]

with disability and allow their social integration with the society on more frequent basis. They would consider the neural prosthetics as a part of their normal life and daily routine. However, it has been observed that the sector is quite promising and rewarding for those individuals or organizations who are willing to work sincerely and devote considerable amount of time, energy and money required to be involved in the field.[8–10]

With the advancements in sensor material technology widespread improvements are possible in the sensory feedback techniques. The only limiting factor could be providing the sensory information to the sensory hardware system. With advancement in sensor material, better resolution and stimulation specificity of nerves could be achieved with only hurdle being the correct stimulation patters to elicit natural and meaningful sensations in sensory nerves which requires correct sensory mapping and appropriate sensor feedback. The requirement to enhance the number of sensory feedback would grow with temperature and limb position although availability of touch, pressure, vibrations, pain and limb movement in the system. Hence, in this area of research it is important to remember that innovation and progress are tied not only to developments in technology but also to the needs of the user. With the overarching goal being to achieve a prosthetic device that can perfectly mimic an intact limb in both control and sensory feedback, there are many steps that remain to get there. The focus should be on improving functionality in the daily lives of individuals living with limb differences or paralysis as shown in Fig. 6.5.

Neuroprosthesis in cognitive and physical domain

The cognitive neural prosthetic (CNP) are quite adaptable which means they can be used to support or assist the paralyzed and dissection patients. Rather than signals strictly related to motor execution or sensation, the CNP records the subject's cognitive state. Neuroprostheses are devices that use electrodes to communicate with the nervous system, intending to restore function lost due to spinal cord damage (SCI). The study of human cognition, as well as the human person as a system, is part of Cognitive Prosthesis. The ultimate goal of cognitive prosthesis is to enhance human capabilities and overcome their limitations based on the information available from the brain. An example of such prosthesis is the Eyeglasses that enlarge but do not replace the eye. Neural prostheses devices could play a vital role to replace a motor, sensory, or cognitive modality that has been lost by an injury or disease. Cochlear implants are one example of such technology.[10]

After developing new perceptions like as human–centered computing, saw a major shift in how we think about information technology and intellectual machines in general. It is an example of a "systems view," in which human thought and action are considered intimately entangled and equally significant parts of analysis, design, and evaluation. This concept is more concerned with computational aids designed to amplify human cognitive (or perception) and perceptual capacities than it is with stand–alone exemplars of mechanical cognitive aptitude. These are cognitive prostheses, intelligent machines, or computing systems that influence and boost human intellectual capacities, similar to how the steam shovel was a muscle prosthesis. The prosthesis metaphor emphasizes the need of creating systems that integrate human and machine components in ways that influence their respective strong point synergistically. Designing the computational prosthesis, need various scientist example: social scientists, cognitive scientists, clinicians, and computer scientists are of various departments, which is a broader interdisciplinary spectrum than has historically been associated with artificial intelligence (AI) work. This modification in perspective emphasizes the importance of human–machine interaction. The "system" shown in question is not "the computer," but also social and cognitive processes, computing system–related tools, as well as physical facilities and the surroundings. As a consequence, human-centered computing offers a renewed outlook on research, as well as new research schemas and objectives.[10–12]

Over the last two decades, neural prosthetics achieved a lot of success in the clinic. DBS (Deep brain stimulation) of the subthalamic nucleus is more effective (successful) than OMC (Optical medical care) in the treatment of Parkinson's disease motor symptoms,[8] while spinal cord stimulation (SCS) depresses pain (>50 percent) in people with persistent back pain and lower leg. Apart from this, the achievement of new medicines and electrode design affords a prospect to improve these therapies.[12] Implantable stimulators which are used in DBS and SCS have a battery life of 3 and 4 years, respectively. As stimulation is typically administered continuously, most patients will require multiple replacement surgeries, which are costly and subject the patient to the risks of surgery, such as infection and hardware complications. As a consequence, simulation efficiency is the one area where the efficiency of stimulation can be enhanced. The second is stimulation selectivity. This is the area that can be upgraded. In DBS and SCS, poor electrode placement is a communal cause of failure. In some cases, lead deviations disregard some or all potential clinical benefits, while in others, malposition causes negative side effects from stimulation of non-target regions. Sometimes small lead misplacements can be addressed by regulating the amplitude, length, and frequency of the applied electrical waveform; however, greater misplacements necessitate a second surgery to relocate the lead.[9,11]

Apart from performance, the dangers associated with electrode insertion and residence must be measured. An electrode implant, shifts brain tissue, causing complete harm to blood tissue, ECM, glia, and neurons. The tissue reacts to the damage and the long-term presence of the electrode by adapting to the environment around it, perhaps leading to more neuron loss. Due to the long-term survival of the targeted neuronal elements, it is most critical for the functioning of the neural prosthesis, both stimulation and recording electrodes must be constructed to minimize tissue response.[12] The essential principles of electrode design for brain stimulation and neural recording are described in this paper, as well as how these ideas might be applied to improve the efficacy of neural prosthetic devices. The design of stimulation electrodes needs to be understood, as well as how they can be made more efficient by consuming less power, more selective by reducing nontarget element coactivation, and less harmful during stimulation. Further, the recording electrodes need to be studied and how they can be made to record more selective neural signals, be less prone to electrical noise, and be less destructive to neural tissue. The future of smart electrode designing is an upcoming and key area of potential new research in the coming years.

Before deploying a neural prosthesis, clinical trials are the last opportunities that are made accessible to patients and potentially by a strong or healthy human. Before starting the medical therapy or treatment the institutional research board must approve all clinical trials and declare them safe for human usage. During clinical trials, the technology's breadth and potential hazards needs to be regularly assessed and determined, so as to avoid harmful effects to patients during use. As a consequence, we must practice extreme caution in clinical studies of neuroprosthetic devices to confirm that participants and anybody with whom they come into contact are safe because this is contagious. Given the wide range of neuroprosthetic devices under investigation, customized safety and ethical requirements for each device should be the norm. When compared to a prosthetic limb or computer interface for a quadriplegic, which may have many significant mechanical components and thus the potential to physically harm the patient or nearby persons, refinements of cochlear implants do not require as much scrutiny because they have a decades-long track record.[13]

Clinical trial participants must be provided with sufficient information to ensure that they have a clear grasp of their involvement and are fully aware of the risks and advantages of participating in clinical trials for investigational devices. They must also be informed about potential drawbacks and benefits for both short- and long-term use. Although it is commonly established that informed consent can never be too informed, neuroprosthesis clinical trials are unique in that even researchers and health care practitioners may not be able to thoroughly assess all potential dangers. This is due to a scarcity of data on prognostic outcomes, particularly for novel neuropros-thetic devices that have only been on the market for a few years. When opposed to pharmacological drugs, there may be less concern about systemic adverse responses from neuroprosthetic devices. Furthermore, restricted physical contact exposure of neuroprosthetic devices may make them safer than investigational medications that interact with bodily tissues system-ically through biochemical processes. Electrochemical interactions at the neuroprosthetics–body interface and shedding of such interface materials, such as carbon nanotube (CNT), are undoubtedly concerning, but they may not trigger systemic reactions as pharmaceutical drugs might.[13]

Clinical trial participants are sometimes highly informed, thanks to rising awareness of emerging technology through mainstream media and the Internet. It would be simple for a disabled person to enroll in a clinical trial that could potentially restore his or her lost bodily function in this information age, but the clinical trial investigator has a moral obligation to make short- and long-term goals, as well as the benefits and drawbacks of enrolling in trials, clear to the patient so that there are no surprises at the end of the trial. Compensation to participants, whether monetary or otherwise, should be elaborated and should not be so high that it interferes disproportionately with their enrolment decision-making process. If there is a potential that the neuroprosthetics will provide participants an unfair advantage, this should be stated explicitly. If there are no obvious constraints on the use of such experimental devices, written and distributed instructions on how to use this technology with good intention and compassion should be developed and distributed to participants. When scientists and engineers are developing and investigating a device or implant, they are typically enthusiastic about the potential benefits of using it. Investigators, on the other hand, have an ethical obligation to fully communicate the risks and benefits of clinical research to patients who are interested in taking part in the study.[13–15]

With widespread usage of the Neural prosthetics or NPs it is interesting that their basic physiological actions are often poorly understood. For example, trains of electrical stimuli applied to a muscle nerve excite both sensory and motor axons. The motor axons activate muscle fibres directly while the sensory input enters the spinal cord where it activates not only local neural circuitry but also ascending pathways. The net result in clinical functional electrical stimulation applications is an interplay between the direct motor and indirect reflex actions. Our incomplete understanding of the mechanisms of electrical stimulation of the nervous system is sometimes seen as a fundamental barrier between those whose primary interests are physiological mechanisms and those involved in developing clinical applications. Yet electrical stimulation can often produce useful clinical and functional outcomes. Hence, it is imperative to have an enhanced scientific understanding for successful clinical device applications which has not progressed much in research discipline. The same could be said of the other clinical approaches discussed at the symposium: regeneration, neuropharmacology and intensive training.[15,16] The brain system is extremely complex and, therefore, the fact that the development of neuroprosthetics trials has often been based on a single discipline may have been a limitation.

The interdisciplinary research in neuroprosthetics involve the medical applications of sensorimotor neuroprosthetics, systems neuroscience and neuroprosthetics and next-generation technologies for neuroprosthetics. The first subcategory, Medical Applications of Sensorimotor Neuroprosthetics, involves the accumulation of clinical studies in patients to investigate future neuroprosthetics. As described above, the most successful neuroprosthetic devices developed to date are cochlear implants for patients with

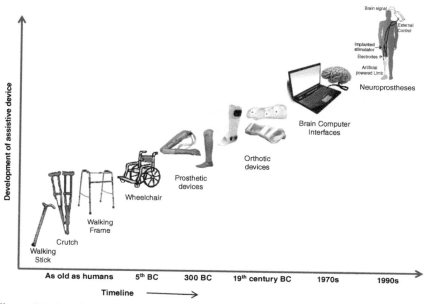

Figure 6.6 Development of various assistive devices with progress of time.[29]

hearing impairment and prosthetic devices for amputees. Therefore, there has been a great deal of research regarding these devices, located on the edge of the sensorimotor processes as shown in Fig. 6.6. However, most of the research discussed here is not specifically focused on simple sensory input or motor output, but rather is related to interactions occurring during sensorimotor processing.

These devices have been developed for paretic patients, for rehabilitation and restoration of lost motor function. With the help of brain–machine interface researchers are able to record signals from brain and connect them to effectors with recording of bio-signals through invasive or non-invasive means as shown in Fig. 6.7. The main area for extraction of motor-related information is the primary cortex area whereas, the signals could be recorded from the middle frontal gyrus region also. This could allow extraction of the complex sensorimotor information from the brain to restore motor function as shown in Fig. 6.8. The patients with could lose voluntary movements and means to communicate because it is closely linked to movement of body parts. The other medical procedures that can be categorized as neuroprosthetics are vagus nerve stimulation or responsive neurostimulation for medically refractive refractory epilepsy and deep brain for movement disorders. Investigations to improve medical practices in this area are currently in progress.

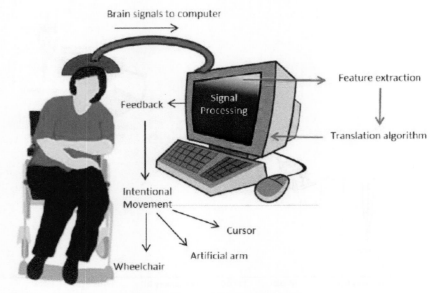

Figure 6.7 Working model of brain computer interface.[29]

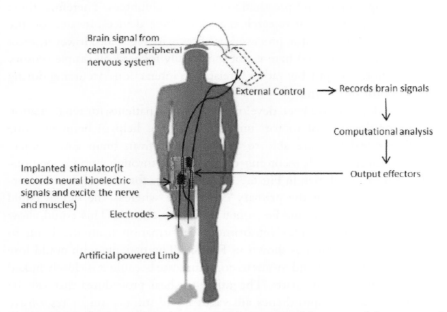

Figure 6.8 Artificial power Limb controlled by processed brain signal from external control an implanted stimulators in *Neuroprostheses*.[29]

Systems Neuroscience and Neuroprosthetics, collect information in healthy individuals or animals and investigate the complex system brain functions. The persons having severely impaired sensory or cognitive functions can employ assistive devices to enhance their perception and functional abilities to perform bodily tasks. With advanced in vivo animal studies it is possible to understand nerve stimulation patterns which induces neuropsychiatric effects. The advanced knowledge from such research studies would allow development of sophisticated neuroprosthetics which could help to treat the nervous system disorders and different types of motor neuronal developmental disorders. With use of advanced engineering technologies or computational modelling it is possible to develop smart devices which could interact with the brain and the outside environment for constant interaction during task performance.

Summary

Neural Prosthetics are designed to reduce the disability of a patient due to damage to neurons. It aims to provide a closed-loop system in which signals from the central nervous system can be sensed and cause the desired motor response and in turn, external signals from the environment can be encrypted and sent to the brain to restore lost sensory information. The nervous system is the body's command center which has two parts: Central Nervous System (CNS) and Peripheral Nervous System (PNS). The CNS consists of the brain and spinal cord whereas the PNS refers to the part outside the brain and the spinal cord.[16]

A neural prosthesis works through a brain–machine interface (BMI), it is a chip implanted in the user's brain where the signal from the brain can be read by the prosthetic device itself. These signals would control an organic limb fire and thus perform the desired function. Every time we think, move, feel or remember something, our neurons are at work. That work is carried out by small electric signals that zip from neurons to neurons as fast as 250 mph. The central nervous system undergoes significant reconstruction. The dynamic change in the interrelation of the brain-spinal cord axis as well as in the structure-function relations plays a crucial role in the determination of neurological functions, which might have important clinical significance for the treatment and evaluation of patients with spinal cord injury.[17]

Motor and sensory dysfunctions after spinal-cord injury (SCI) result in a functional reorganization sensorimotor network where the dorsal column transmits sensory information to the brain, and the anterior column is exclusively transmitting motor commands to the ventral horn motor neurons. The lateral column, however, transfers both sensory and motor information between the spinal cord and brain. It also involves somatosensory which helps to interpret bodily sensation. The sensorimotor network is responsible for sensing physical inputs, converting them to electrical signals that travel throughout the brain network, and then initiating a physical response.[18,19]

Ground-breaking developments are being done in the field of motor neuroprosthetics where a device interprets the brainwaves from the motor cortex into electrical signals that are transmitted to a computer, which in turn connected to a motor device to move the limbs of the patient. This is the application for Spinal–cord injury, limb loss, and neuromuscular disorders patients where a communication board, a wheelchair, or prosthesis is allowed for operation.[19–21] The majority of patients with hearing loss notable enough to result in social dysfunction can be treated with non-surgical interventions. Cochlear device implantation (CDI) remains the only option for auditory communication rehabilitation in cases of severe and intense sensorineural and hearing loss where the site of lesion is outside of the central auditory processing stream. The healthy normal–functioning human cochlea can hear the sound signals in the frequency range of 20 kHz, in the basal region, to 20 Hz in the apical region. Electrode arrays are designed with particular characteristics that allow for the protection of intra–cochlear structures during the insertion process, as well as during explanation. Two types are available: Straight lateral wall and pre-curved modular hugging. The 'insertion depth of an electrode depends on the length of cochlea dust. However the electrode array should not cause any degree of trauma to any of the intra–cochlear structures during insertion, also the surgeons should operate without any complications. Cochlear implantation is becoming the most successful neural prosthesis in clinical use and its long history of innovations has resulted in high-performing devices.[22–24]

Cognitive Neural Prosthetics (CNP) seeks to help paralyzed patients by recording their thoughts directly from the brain and decoding them to control external devices such as computer interfaces, robotic limbs, and muscle stimulators. Spinal cord lesion and other traumatic accidents, peripheral neuropathies, amyotrophic lateral sclerosis, and stroke can result in paralysis.[25] Electrical stimulation can produce improved clinical and functional outcomes. Further, the regeneration, neuropharmacology and intensive patient training of spinal cord injured patients can show better results and successful clinical applications. Hence, all round management of patients is required to achieve desired results in cognitive enhancement and better patient care. The benefits of closed feedback systems depends on task complexity. The simple tasks could be easily performed however, demanding tasks immensely benefit from feedback control through feedback sources such as sound and vision. The user experience is critical but complex and varies from individual to individual. Once user gets familiarised with the task, they may improve their feedforward performance, so that feedback becomes more or less redundant. However, certain users showed opposite trend wherein to control a prosthetic device, they immensely benefitted from feedback learning process. They found it useful, comfortable, friendly and as part of their routine daily life. The benefit from feedback could only be experienced after a certain adaptation period during which user learning would enhance and they would understand how to exploit the device to it maximum optimum utilization. Future research should focus on investigating the interactions between task learning and feedback to understand the relationship of each feedback variable such as touch, force and degrees of freedom (DOF) and how they contribute to overall task performance. Hence, the role and benefit of feedback is indeed quite complex and dependent on several factors some of which could vary. They need to be understood well before designing effective feedback controlled prosthetic devices. Further, a multi-functional coding system on a limited number of vibrotactile motors could possibly make an ideal solution to be employed in advanced limb prosthetics.[25,26]

The future neural prosthetics would help in treatment of epileptic seizures, chronic pain, Traumatic brain injury (TBI) patients, people with sensory and motor disabilities.[27,28] such as Alzheimer or Parkinson disease, speech disabilities or people with limb malfunction due to injury or war. Research is ongoing with several applications in these fields may soon find market. The role of engineers to design effective and innovative materials that are biocompatible that allow design and fabrication of smart devices which may be tolerant to ambient environmental conditions without causing damage or adverse reaction to any body organ. Therefore, potential need to develop suitable material and interface technologies with availability of micro-devices that can be safely implanted inside the human body for long time duration is utmost needed. Such technology is already being employed in auditory prosthetic applications wherein the material can mimic the neural tissue more closely and work better than presently available cochlear or deep brain stimulating devices. It is found that polymer is more compatible to human body than silicon which was conventionally used in neural probes. This may avoid damage and tearing of the tissue and handle the delicate parts carefully without adversely affecting the other body functions. The disadvantage of being flexible may cause difficulty in fixing the implant into the body. Hence, the scope of employing neural prosthetics is quite wide and enormous but it depends on the scientific community on how to use them for design and fabrication of advanced devices which could help in providing promising solutions to the patients and their families.[25,28,29]

References

1. Patel S, Park H, Bonato P, Chan L, Rodges M. A review of wearable sensors and systems with application in rehabilitation. *J Neuro-eng Rehabil*. 2012;9(21):1–17.
2. Loeb GeE. Neural Prosthetics: a Review of Empirical vs. Systems Engineering Strategies. *Appl Bionics Biomech*. 2018;1435030:1–17.
3. Prochzka A, Vivian K, Mushawarab Douglas B, McCreery. *Neural Prosthesis J Physiol*. 2001;533(1):99–109.
4. Wilson L, Horton L, Kunzmann K, et al. Understanding the relationship between cognitive performance and function in daily life after traumatic brain injury. *J Neurol Neurosurg Psychiatry*. 2021;92:407–417.
5. Dresler M, Sandberg A, Bublitz C, Ohla K et al. 2019, Hacking the Brain: dimensions of Cognitive Enhancement, 10, pp. 1137-1148.
6. Rosenfeld JeV, Bandopadhayay P, Brown DoJ. Top Spinal cord Injury Rehabilitation. *Summer*. 2008;14(1):76–88.
7. Maas AIR, Menon DK, Adelson PD, et al. Traumatic brain injury: integrated approaches to improve prevention, clinical care, and research. *The Lancet Neurology*. 2017;16:987–1048.
8. Wilson L, Stewart W, Dams-O'Connor K, et al. The chronic and evolving neurological consequences of traumatic brain injury. *The Lancet Neurology*. 2017;16:813–825.
9. Andelic N, Løvstad M, Norup A, et al. Editorial: impact of traumatic brain injuries on participation in daily life and work: recent research and future directions. *Front Neurol*. 2019;10:1153.
10. Chaytor N, Temkin N, Machamer J, et al. The ecological validity of neuropsychological assessment and the role of depressive symptoms in moderate to severe traumatic brain injury. *J Int Neuropsychol Soc*. 2007;13:377–385.
11. Mushahwar VK, Horch KW. Muscle recruitment through electrical stimulation of the lumbosacral spinal cord. *IEEE Trans Rehabil Eng*. 2000;8:22–29.
12. Ramon-Cueto A, Cordero MI, Santos-Benito FF, Avila J. Functional recovery of paraplegic rats and motor axon regeneration in their spinal cords by olfactory ensheathing glia. *Neuron*. 2000;25:425–435.
13. Hansen GVO. EMG-controlled functional electrical stimulation of the paretic hand. *Scand J Rehabil Med*. 1979;11:189–193.
14. Jones LL, Oudega M, Bartlett Bunge M, Tuszyski MH. Neurotrophic factors, cellular bridges, and gene therapy for spinal cord injury. *Physiol J*. 2001;533:83–89.
15. Taub E. Constraint-induced movement therapy and massed practice. *Stroke*. 2000;31:986–988.
16. Taylor PN, Burridge JH, Dunkerley AL, Wood DE, Norton JA, Singleton C, et al. Clinical use of the Odstock dropped foot stimulator. Its effect on the speed and effort of walking. *Arch Phys Med Rehabil*. 1999;80:1577–1583.
17. Weingarden HP, Zeilig G, Heruti R, Shemesh Y, et al. Hybrid functional electrical stimulation orthosis system for the upper limb: effects on spasticity in chronic stable hemiplegia. *Am J Phys Med Rehabil*. 1998;77:276–281.

18. Fletcher MD, Tini N, Perry SW. Enhanced pitch discrimination for cochlear implant users with a new haptic neuroprosthetic. *Sci Rep.* 2020;10:10354.
19. Toyama S, Takano K, Kansaku K. A non-adhesive solid–gel electrode for a non-invasive brain-machine interface. *Front Neurol.* 2012;3:114.
20. Waltz JM. Spinal cord stimulation: a quarter century of development and investigation. A review of its development and effectiveness in 1,336 cases. *Stereotact Funct Neurosurg.* 1997;69:288–299.
21. Wheeler Jr JS, Walter JW. Acute urologic management of the patient with spinal cord injury. *Urol Clin North Am.* 1995;20:403–412.
22. Gupta M, Rajak BL, Bhatia D, Mukherjee A. Neuromodulatory effect of repetitive transcranial magnetic stimulation pulses on functional motor performances of spastic cerebral palsy children. *J Med Eng Technol. Taylor and Francis.* 2018;42(5):352–358.
23. Gupta M, Bhatia D. Study the Cognitive Changes in Cerebral Palsy Children Employing Repetitive Transcranial Magnetic Stimulation and Neurofeedback Training. *Lecture Notes in Networks and Systems- Proceedings of the International Conference on Computing and Communication Systems*; 2021:431–439.
24. Gupta M, Bhatia D. *Neurofeedback: Retrain the Brain*, 2, IGI Global Publications; 2019:13–25.
25. Avutu SR, Bhatia D. *Neurological Disorders, Rehabilitation, and Associated Technologies: An Overview*, 3, IGI Global Publications; 2019:26–38.
26. Avutu SR, Paul S, Bhatia D. Smart Rehabilitation for Neuro-Disability: a Review. In: Paul S, ed. *Application of Biomedical Engineering in Neuroscience*. Singapore: Springer; 2019:477–490.
27. Gupta M, Dinesh B. Retrain the Brain through Noninvasive Medically Acclaimed Instruments. In: Paul S, ed. *Application of Biomedical Engineering in Neuroscience*. Singapore: Springer; 2019:51–60.
28. Bhatia D, Wearable technological innovations in Neurosurgery Progress in Clinical Neurosciences. vol. 34. Thieme Publishers; 2021. pp. 87–92. ISBN: 978-93-90553-94-5.
29. Rajak BL, Gupta M, Bhatia D. Growth and advancement in Neural control of limb. *Biomed Sci Eng.* 2015;3(3):46–64.

CHAPTER 7

Virtual reality, augmented reality technologies, and rehabilitation

Introduction

Virtual reality, BCI, and Robotics are innovations, now a days used in physical rehabilitation to work with and supplement conventional rehabilitation techniques. A robot can be characterized as "a machine that performs different complex demonstrations of a person" and "naturally performs confounded, repetitive movements." Current robotic frameworks incorporate remedial robots, for example, large fixed exercise robots or wearable exoskeletal robots. There are likewise utilitarian robots like assistance robots for exercises of daily living (ADL), and companion robots.[1] Virtual reality (VR) is characterized as a human-computer interface that permits a client to associate with a computer-produced environment, utilizing different sensory channels.[2] The virtual environment (VE) can assist clients with visual, auditory, or tactile criticism. Current frameworks fluctuate in size, scope, and level of submersion, from cave programmed virtual environments, which are room-sized establishments containing 3D visual and auditory frameworks, to head-mounted visual displays (HMD).[3]

Augmented reality (AR) adds advanced components to a live view through a gadget generally a smartphone, tablet, or wearable glasses. Models incorporate projector-based portrayal of appendages for handicapped people and multi-faceted maps with superimposed names and data. Not at all like VR which establishes a different environment, AR adds a computerized layer to the real world. Various sorts of AR exist on a continuum from unaltered reality to a completely virtual environment. An assortment of movement following sensors or cameras have been created to consolidate client movement and position. New virtual reality applications currently use cell phones as both an intuitive visual climate and furthermore an information recording gadget, permitting estimations of movement and function to be gathered in the virtual climate, making the innovation more adaptable and accessible.

Modern Intervention Tools for Rehabilitation.
DOI: https://doi.org/10.1016/B978-0-323-99124-7.00001-8

In the field of physical rehabilitation, robotics and VR frameworks can possibly measure capacities, just as to fill in as therapeutic tools. These frameworks can be utilized to evaluate and measure motor capacities, posture and limb position, strength, gait, and equilibrium. Real-time feedback in regards to execution can be given to patients and therapists. Reports can likewise be created to give objective information in regards to patient progress in contrast with themselves or different clients. Therapeutically, they assist clients with repetitive, contextualized, task-explicit preparing in a stimulating and engaging way. These kinds of activities are essential in neurorehabilitation as they help in promoting neuroplasticity and patient recovery.[4] Moreover, the engaging presentation of the generally redundant or ordinary tasks with game-like interfaces can help with compliance of the patient with treatment programs after some time.

Moreover, robotic frameworks can work with or help the movement of paretic limbs in performing useful movements and activities, empowering patients to build social commitment and increased independence and autonomy. VR frameworks can give cognitive activities and measures just as a re-enactment of real-world activities in a protected setting.[5]

Robotic frameworks are interestingly fit to furnish functional assistance with mobility and daily living activities. For instance, robotic exoskeletons can permit paraplegic people with SCI to walk independently. Brain-computer interfaces, where computers investigate mind cues and use them to control robotic devices are being read up for patients with neuromuscular issues, for example, amyotrophic lateral sclerosis.[6] Brain-computer interfaces (BCI) empower clients to change their thoughts into activities without including intentional motor movement. This somewhat new UI innovation not just offers those with extreme neuromotor inabilities another means to associate with the environment, however can likewise give viable and engaging rehabilitation to re-establish motor or cognitive capacities that have been impeded because of any injury or trauma.[7]

VR frameworks are likewise being utilized in the field of orthopedics and sports medication to give a drawing-in, organized environment for performing strengthening, stretching, cardiovascular exercise, and for prosthetic training. Also, VR frameworks have been utilized to treat cognitive problems coming about because of CNS disorders. Repetitive, high-intensity task-explicit preparing has been displayed to spike neuroplasticity and functional recovery post-stroke. Commonly this kind of treatment is accomplished by a group of doctors and specialists. This methodology is both time-concentrated and costly in terms of assets and skilled therapists. Hypothetically VR and robotics can automate segments of this process, making it more reproducible, information-driven, and financially effective. Robotics and VR can be used anytime during the process of rehabilitation, from the acute to chronic stages. These advances might be utilized to enhance traditional treatment to accomplish the applicable objectives at each phase of recuperation and recovery. evidence with respect to advanced mechanics and VR is still too restricted to even think about giving authoritative clinical practice suggestions, and it stays hazy whether its incorporation is more valuable at one phase of rehabilitation than others. Wearable powered orthoses can serve both practical and restorative objectives by making up for a patient's neurological deficiencies while preparing muscles and coordination. Stroke patients utilizing an electromyogram (EMG)- controlled exoskeletal upper limb powered orthosis showed improvement in function and spasticity.[8]

In research of controlled exoskeletal robots for the lower limbs in patients with motor complete SCI, exoskeletal robots were demonstrated to be safe and very much endured, and members had the option to walk freely for the time of five to ten minutes were reported with physical and psychological advantages.[9] Notwithstanding the motor and sensory training, VR can likewise be utilized to evaluate and perform cognitive activities to deal with executive work, memory, attention, arranging, and visuospatial handling.[2] The innovation additionally has been examined in different recreation settings with virtual workplaces, kitchens, and driving conditions. Inside the pediatric populace, VR has been utilized among kids with CP and was viewed as viable in further developing engagement and motor work, proximal stability, postural tone, and range of motion.[10,11]

Virtual reality

Virtual Reality portrays a computer-created situation (a virtual world) with which the client can connect in 3 dimensions so the patient feels that the individual is part of the scene.[12] VR can be portrayed as "an advanced type of human-computer interface that permits the client to connect with and become submerged in a computer produced naturalistic environment.[13] For its adaptability, a feeling of presence (i.e., the sensation of "being there"), and emotional engagement, VR rehabilitation has been tried and tested in individuals with motor and cognitive deficits, resulting in great outcomes. In stroke patients, the quantity of VR programs is quickly expanding with convincing information showing an improvement in the recuperation of motor capacity and activities of daily living.[14] VR permits a degree of engagement and cognitive contribution, higher than the one given by memory and creative mind, however is more controlled and can be more easily estimated than that presented by direct "real" experience.[15] Its multisensory stimulation implies VR can be viewed as an enhanced environment that can offer functional and natural real-world requests[16] (e.g., tracking down objects, gathering things, and purchasing stuff) that might further develop brain plasticity and regenerative processes.[17] There are a few models in the writing where VR has been effectively utilized both as an assessment tool and as a therapeutic tool.[18] As an evaluation tool, VR has been utilized to distinguish visual-vestibular deficits in grown-ups after a concussion and mild TBI.[19] At present, there are 4 types of virtual environments: head-mounted display, augmented, Fish Tank, and projection-based.[20] An absolutely immersive VR framework is the head-mounted display (HMD) where the subject sees just the computer-created picture and the remainder of the actual world is obstructed from view. With augmented VR frameworks, both computer-produced pictures and the actual world are apparent to the subject. Subsequently, the PC world is overlaid on the physical world. With "Fish Tank" VR, the sound system pictures are delivered on a screen before the subject.[21] These frameworks have a restricted field of view (FOV) and space in which one can associate with the scene. Thus, the subsequent FOV is smaller than that accessible with other VR frameworks yet the going with pixel visual point is additionally smaller and, in this way, better. With projection-based VR, the computer-produced imagery is projected on a screen or wall before the client similar to that in a theatre.[22] Back-projection is regularly utilized rather than front-projection to safeguard that the projected scene isn't darkened by the subject's body. These frameworks normally have a wide field of view and can be multi-walled and floor frameworks likewise to the CAVE[TM] innovation.

Many individuals question why we don't simply have subjects perform motor tasks in reality. The response to this inquiry is that VR offers us the potential chance to bring the intricacy of the actual world into the controlled climate of the research facility. VR gives us the possibility to get away from reductionism in science and towards the estimation of natural development inside normal complex conditions. As a general rule, VR permits us to establish an engineered climate with exact command over countless physical factors that impact conduct while recording physiological and kinematic reactions.[23] There are various qualities underlying the utilization of VR with rehabilitation.[24,25] Among these are that VR gives the opportunity for ecological legitimacy, stimulus control, and consistency, real-time performance feedback, autonomous practice, upgrade and reaction changes that are dependent upon a client's actual capacities, protected testing and preparing the environment, the chance for graduated exposure to stimuli, the capacity to distract or augment the performer's attention, and maybe generally essential to therapeutic intercession, motivation for the client.

As a therapeutic instrument, Dahdah et al. demonstrated that vivid VR intercession can be utilized as a successful neuro-rehabilitative apparatus to improve executive capacities and data handling in the sub-acute period, giving proof of beneficial outcomes of a virtual Stroop task over customary non-VR-based task.[26] VR as a therapeutic tool utilized for training attention in serious traumatic brain injury (TBI) with positive outcomes in the early recovery stages[27] with a particular "augmented" task in which virtual and haptic criticisms were utilized in a target-reaching exercise to improve sustained attention. At last, virtual protocols created upon commercially accessible game arrangements have been compelling in addressing and treating balance deficits.[28] Besnard et al. made a virtual kitchen to survey day-to-day activities and assess chief dysfunctions in subjects with extreme TBI.[29] Robitaille et al. fostered a VR avatar interaction stage to evaluate remaining executive functions in clients with mild TBI. The platform can capture the real-time movement of the subject translating them into a virtual body, that is, hence positioned in a simulated environment (i.e., a town).[30] The client is then permitted to investigate the simulated environmental elements which include diverse navigational deterrents to overcome. Comparable methodologies have been utilized by different authors, though simplified settings (i.e., 3D virtual passage that the subject can be investigated with a joystick) have been demonstrated helpful to evaluate subclinical cognitive anomalies in asymptomatic subjects that experienced a concussion.[31] This large number of works recommend that VR could be helpful as an evaluation tool and in rehabilitation. Alternately, VR treatment conventions for cognitive recovery are utilized transitionally from mild to extreme conditions, in spite of the fact that the viability of these sorts of intercessions should be additionally investigated.[32]

Augmented reality

The term Augmented reality (AR) comprehensively portrays processing devices that overlay computerized data onto a perspective of the physical world. While there are many types of AR, including smartphone or tablet-based showcases, head-mounted 3D AR (3D-AR) gadgets present exciting additional opportunities for clinical interventions. 3D-AR gadgets are head-worn processing gadgets that show virtual objects like they are available in the client's physical world. 3D-AR gadgets usually have two transparent displays that demonstrate in front of the client's eyes. Utilizing stereoscopic delivering with the two screens, the gadgets show advanced content that shows up as 3D hologram images to the client. 3D-AR gadgets commonly join on-board spatial tracking hardware and calculations that track the headset's movement in the actual climate. The spatial tracking permits 3D-AR gadgets to show multi-dimensional images that show up in stable locations in clients' surroundings, which makes the 3D experience more vivid. Numerous AR innovation-based exploration models and business items have been accounted for in the use of image-guided treatment.

Commercial AR products in image-guided interventions

Numerous business AR items are accessible on the market. Among these, Microsoft HoloLens is an AR item that has been utilized in a wide scope of clinical applications.[33] The HoloLens is a head-worn showcase (HWD) intended to project data by means of three-dimensional (3D) images with different sensors.[34] To decide its position in space, HoloLens is furnished with an inertial estimation unit, depth detecting camera with infra-red (IR) sensor, and RGB camera. HoloLens is additionally ready to get client input through voice orders and hand motions.[35] Other than HoloLens, there are other business AR items applied in image-guided treatment, like Endosight and Magic Leap. The Endosight is explicitly intended for biopsy and ablation procedures. The current variant of the item incorporates a dispensable optical sensor to be inserted in third-party probes or hardware.[36] The real-time AR picture from Endosight can be seen from AR goggles or tablets in color.[37] The Magic Leap is more like HoloLens, it is not explicitly intended for clinical application only but the utilization of it in image-guided clinical strategies.

Mojica et al.[38] created and evaluated an interface for image-guided preparation of prostate intercessions with the utilization of a head-mounted display (HMD). Created with Unity3D Engine and written in C#, the point of interaction was tried on HoloLens and required the utilization of hand motions and voice orders to control pictures and holographic scenes.

Kuzhagaliyev et al. introduced an AR framework that focuses on working on the proficiency of irreversible electroporation (IRE) by showing a point of passage to the body through a drawing of the current needle counterbalance and the objective trajectory with HoloLens holographic guides. The framework was created utilizing the Universal Windows Platforms (UWP) Unity3D engine. Park et al.[39] assessed a HoloLens based 3D AR-assisted navigation framework by utilizing CT-directed simulations. The AR framework was a HoloLens headset-based framework and the perception and interaction applications were created through Unity and Mixed Reality Toolkit Foundation. The created framework could accomplish computerized enlistment between the 3D model and CT grid through PC vision and Vuforia with next to no client input.

Solbiati et al.[40] planned a progression of analyses to assess the attainability of the novel Endosight AR framework. The Endosight framework has a modified needle handle, radiopaque labels as fiducial markers, and a tablet used to run the Endosight programming and show the AR pictures. The work process of the Endosight framework begins with fiducial markers being put on the patient's skin before CT checking. Then, at that point, Endosight programming tracks down the target tumor and fiducial markers on the CT pictures naturally by means of a division algorithm. Then, during the intercession, the pre-intervention data is attached with real-time data from the needle handle and fiducial markers to produce real-time AR pictures on the tablet. Lastly, radiologists finish the intercession in light of the real-time AR pictures.

Brain–computer interface

Brain-computer interface (BCI) innovation gives novel neuro-engineering answers for restoration issues brought about by amputation or any other neurological deficit. Thus, neural communicating procedures are being incorporated in recovery techniques across patient populations. Brain-computer interface (BCI), likewise alluded to as a brain-machine interface (BMI), is a hardware and programming framework that empowers people to connect with their environmental elements, without the inclusion of peripheral nerves and muscles, by utilizing control signals produced from electroencephalographic activity. BCI creates another non-muscular channel for relaying an individual's intentions to outside gadgets like computers, speech synthesizers, assistive machines, and neural prostheses. That is especially alluring for person having serious motor incapacities. Such an interface would further develop their personal satisfaction and would, simultaneously, decrease the expense of intensive care.

A BCI is an artificial intelligence framework that can perceive a specific arrangement of patterns in brain cues following five sequential stages: signal acquisition, pre-processing or signal improvement, feature extraction, classification, and the control interface.[41] The signal acquisition stage catches the mind cues and may likewise perform noise reduction and artifact handling. The pre-processing stage readies the signs in a reasonable structure for additional handling. The feature extraction stage distinguishes discriminative data from the recorded brain signals. When estimated, the signals are planned onto a vector containing effective and discriminant highlights from the noticed signal. The extraction of this fascinating data is an exceptionally difficult errand. The classification stage characterizes the signals considering the element vectors. The decision of good discriminative elements is in this manner fundamental to accomplish effective pattern recognition, to interpret the client's intention. At last, the control interface stage makes an interpretation of the arranged signs into significant commands for any associated device, like a wheelchair or a PC.

BCI technology

The centre elements of the latest BCI frameworks are recording neural signs from the brain, processing these signs through a computer algorithm, and making an interpretation of handled signs into a planned activity utilizing an end-effector device.

Recording and decoding

While one could hypothetically get to any level of the central sensory system with BCI innovation, a few regions give reasonable benefits over others. The cerebral cortex has practical geography that empowers relatively simple focusing of the explicit motor and tactile subsystem. For instance, the hand region of the primary motor cortex has an unmistakable anatomical morphology that is additionally conserved across people and is strategically placed on the lateral aspect of the cerebral convexity, which allows easy targeting of either invasive or non-invasive frameworks.[42] Additionally, the spinal cord has a geography that is grossly saved across people and takes into account recording and stimulation of the motor corticospinal tracts and sensory dorsal columns.[43] A critical pragmatic differentiator in neural recording advances is the degree of intrusiveness of a particular strategy, usually classified as "invasive" or "non-invasive," depending upon whether placement requires penetrating the integument. Non-invasive recording procedures incorporate direct proportions of electrical movement coming from neuron depolarization, for example, electroencephalography (EEG) recordings, or indirect proportions of neuron firing, for example, functional magnetic resonance imaging, magnetoencephalography,[44] or functional near-infrared spectroscopy.[45] Rather than non-invasive procedures, invasive recording strategies present the advantages of more noteworthy spatial and temporal specificity yet bring about the additional dangers of surgically implantable gadgets. However, there are fluctuating levels of obtrusiveness.[46] In the least-intrusive degree of recording strategies, the electrode are placed just beneath the skin surface, diminishing the separation from the nervous system to the recording electrode and working on signal quality.[47] Alternatively, electrodes can be set beneath the hard bony structures in the epidural or subdural space, lying as a net of electrodes over neural tissue of interest, as in electrocorticography (ECoG). In the most intrusive case, multiunit microarrays of 100þ electrodes can be embedded into the neural parenchyma to record straightforwardly from individual neurons or little neuronal populaces. Notwithstanding the recording strategy, when crude neural signals are obtained, they are handled with amplification and digitization. Key elements of the digitized neural signals incorporate both amplitude and spike train frequency.[48] Remarkable elements are then separated utilizing an algorithm and converted into device commands. These relevant, extractable elements vary on the neural recording strategy, with harmless methods like EEG utilizing signals from a wide neuronal populace and intrusive techniques depending on individual (or group of) neurons. Besides, neural signal features might adjust to the successes and mistakes of the client utilizing the BCI and might be impacted by physical stimulus (eg, pain) and brain states (eg, weakness or frustration) giving rise to the extra noise to the framework. Neural interpreting algorithms are unique to every BCI and should be versatile, obliging to new clients, and should have the ability to self-adjust this shared variation of both "the client to the framework and the framework to the client" is unique to BCI.[49]

Neural features and degrees of freedom

The number and nature of autonomous, notable highlights ready to be extricated straightforwardly impact the degree of freedom for the end effector. While controlling a computer mouse or mechanical arm normally requires 2–4degrees of freedom,[50] accomplishing dexterity of the singular joints of the hand could need as numerous as 22degrees of freedom. Less-intrusive frameworks, for example, EEG function admirably for a restricted degree of freedom applications, in spite of the fact that extracting countless control signals from the averaged neuronal populace becomes troublesome and conceivably restricts the versatility of such procedures. Intrusive frameworks permit a more prominent number of control signs to be obtained, compared to a more noteworthy degree of freedom for end effectors and making embedded BCIs bound to be fruitful for complex errands.[51] Furthermore, invasive BCI frameworks permit people to practice naturalistic control[52] of end-effector gadgets, implying that as opposed to envisioning[53] or participating in proxy developments of different appendages,[54] as is needed in EEG-based or myoelectric control frameworks,[55] intracortical and ECoG BCIs give the capacity[56] to precisely interpret motor imagery for the real intended movement.[57,58]

External output devices

End-effector targets are the external output gadgets working through BCI commands.[59] These gadgets differ in activity and configuration depending on the motivation behind the BCI framework just as the requirements of each end-user.[60] Coordinating BCI with end–effector gadgets offers difficulties to clinicians and engineers the same.[61] Specialized difficulties incorporate acquiring steady, long haul accounts of huge neuronal assemblies from various regions of the brain, creating computationally proficient algorithms that can change neuronal action into command signals that can handle end effector gadgets, creating shared control to permit ideal association among the patient's BCI control and the pre-programmed control of an end effector gadget, and working in flexibility that can alter calculations as a client's capacities and execution change over time.[62] Here we will focus on development control and neurorehabilitation as end-effector targets of BCIs, including functional electrical stimulation (FES), orthotics and exoskeletons, or prosthetic and robotic devices.

Neuromuscular FES

BCI frameworks combined with FES can work with purposeful control of movements.[63,64] Thusly, BCI-FES might be utilized as an assistive gadget to expand or replace volitional muscle constriction, returning practically useful movement to an incapacitated person's limb (e.g., empowering object manipulation[65] or ambulation).[66] Assistive FES frameworks might be embedded, like the Freehand framework and its later manifestations,[67,68] or transcutaneous[69] with enormous electrode pads[70] or complex varieties of more modest electrodes.[71] Since FES isn't consistently fruitful in creating solidarity to help paralyzed limbs against gravity,[72] a few FES frameworks are created as mixtures with an exoskeleton[73] or splint-like parts.[74] FES can be utilized restoratively to condition muscles,[75] lessen spasticity,[76] or advance neuro recovery from paralysis.[77] When joined with a BCI, evoked FES is put under client control, which might engage[78] with cooperative learning and build up neuromuscular coactivation designs. One illustration of this is increased dorsiflexion and electro-myography activity in the tibialis anterior with BCI-FES as contrasted with FES without BCI in an individual with a stroke.[79]

There are admonitions to this methodology of an end effector for BCIs. To begin with, adding BCI control to FES innovation adds one more element of trouble to the designing difficulties previously presented by utilizing FES to evoke complicated, coordinated movements. Besides, FES may not be the best methodology for all end-user clients. For instance, the muscle that has been denervated by lower motor neuron disease or peripheral nerve injury commonly doesn't react to FES. Additionally, few people with tactile hyperesthesia may not endure electrical feeling at the power that causes muscle constriction. In these cases, movement of the impacted limb might be better accomplished through exoskeletons or robots. In instances of removal or loss of somatosensation, where the ideal remedial impact is the subjective experience of moving the impacted limb, e.g., to alleviate pain states or connect with sensory-dependent plasticity, virtual reality (VR) might be the better end effector.

Virtual reality and BCI

VR is being examined as an end effector for BCI, especially applied to upper limb rehabilitation[80] in stroke.[81] or on the other hand for preparing and assessment of imagery-based deciphering.[82] One advantage of combining

BCI with a virtual environment might be to build patient engagement[83] in retraining through improved visual or potentially haptic feedback.[84] BCI-control in VR framework might assist with engagement of motor learning through repetitive training, in any event, when the client has the restricted capacity to move their limb.

Orthoses, exoskeleton, and robots

Exoskeletons and orthotics are helpful for static bracing of joints into their functional positions and furthermore to create a range of motion movement of the joints. This methodology is useful for patients with lower motor neuron disease or muscle atrophy, where FES is probably not going to evoke usable power. There is likewise developing interest in applying BCI with exoskeleton gadgets in people with upper motor neuropathology, like stroke and spinal cord injuries. powered exoskeletons can increase deliberate and FES-evoked muscle strength and support weak or unstable joints adequately to take into account treatment cooperation. Both upper and lower appendage orthoses have been joined with BCIs as assistive gadgets (e.g., grasp orthosis,[85] Rex Bionics or the HAL exoskeleton for walking)[86] or treatment helps (e.g., BOTAS[87] for upper limb, and ReWalk Robotics, and Ekso Bionics for lower appendage restoration) a portion of these end effectors (eg, ReWalk) has gotten leeway from the Food and Drug Administration as assistive gadgets or treatment aids.

Rehabilitation through robots may augment in the upcoming years due to their capacity of delivering high-intensity, repetitive exercise training to the individual, this element when combined with BCI'S role in the rehabilitation, makes an integrated system of BCI-robot set up which gives adequate support for the motor retraining. It offers the chance of consolidating 2 innovations that can improve patient engagement and gain by the possibility to expand the intensity of motor practice.

The potential of BCI interface to change the recovery of neurologic conditions has brought about an expanded joint effort between biomedical specialists and clinicians. The application of BCI end-effector interfaces in the clinical rehabilitation setting is untimely, there have been clear advances toward this combination. Enhancements in BCI innovation can permit generally quick investigation of intricate neural data and empower estimation of neural signs that can undoubtedly be changed over to control signals. One more interesting change in what's to come is hybrid BCI, which involves the utilization of physiological signals other than EEG as information sources.[88]

BCI applications in rehabilitation

Two critical jobs of BCIs in recovery are replacement and restoration of lost neurologic capacity. At the point when BCI frameworks are utilized to supplant lost neurologic capacity, the innovation re-establishes the client's capacity to communicate with and control different conditions and exercises, including computer-based activities (e.g., word processing, Internet perusing, and so on), environmental control units (e.g., light, heat, TV, and so forth), mobility devices (e.g., power wheelchair[89] drive), or neuroprosthetic limbs[90,91] and orthoses.[92] On the other hand, Brain-Computer Interfaces in Rehabilitation BCIs can be involved with rehabilitative treatments[93] with an end goal to assist with re-establishing normal central nervous system work by initiating activity-dependent plasticity of the brain.[94] These BCI frameworks synchronize brain activity that is similar to movement intent with genuine movements and sensations produced by end-effector gadgets.

It is muddled what minimal neurologic capacity is vital for a BCI model to be possibly effective. For instance, BCIs have been utilized as augmentative specialized gadgets for people with acquired brain injury causing aphasia,[95] locked-in syndrome,[96] and disorders associated with consciousness.[97] However, motor and sensory BCI frameworks utilizing implanted cathodes have depended on intact brain-substrate in precentral and postcentral gyrus regions to catch or animate neural signs through the BCI.[98]

There might be some adaptability in applicant recording regions[99] for BCI-intervened motor control,[100] with supplementary motor regions[101] and parietal sensory regions[102,103] as choices if primary motor cortex is inaccessible because of upper motor neuron illness or injury.[104] There is proof that those with corticospinal tracts harmed by stroke[105] can operate motor BCI frameworks for remedial gains[106] Less is known about significant awareness of whether surrogate cortical locales can be utilized for sensory restoration.

Motor restoration (spinal cord injury, stroke)

Motor BCI frameworks have been utilized as investigational assistive gadgets[107] for people with ongoing loss of motion[108] from different causes,[109] including cervical spinal cord injury spinocerebellar degeneration,[110] amyotrophic lateral sclerosis,[111,112] and brainstem stroke. These frameworks have been combined with surface FES, embedded FES, hybrid FES, exoskeletons, orthotics, upper limb animated avatars, and automated arms.[113]

EEG-based motor BCI frameworks have been utilized to control grasp orthotics, with EEG signals utilized as a chance to progress between development states (on versus off, palmar versus parallel grasp) or cycle through steps in a development succession. Hybrid EEG-BCI frameworks have shown the capacity to control multi-joint gadgets by consolidating other biological signs, e.g., joint position sensors or binocular eye-tracking GPS devices. For instance, clients of an elbow-wrist exoskeleton and FES grasp orthotic can choose grasp, elbow flexion, or rest capacities utilizing EEG signals and control grasp and flexion through a shoulder sensor. Then again, arriving at the direction of a 7-degree of-freedom automated arm can be determined from the stereoscopic gaze, with EEG used to trigger the reach sequence. Thusly, people with tetraparesis have accomplished simple practical functional control with EEG-based BCIs and embedded grasp neuroprosthetic,[114] orthotics, hybrid FES-exoskeleton gadgets, and robotic limbs. Embedded BCI frameworks have been utilized by people with tetraparesis to perform skilled object-control tasks, either with an automated arm or transcutaneous FES.[115] Rather than venturing through a predetermined pattern of hand states, intracortical BCI frameworks give the capacity to choose between numerous FES-evoked hand states or give relative control of the multi-joint range of motion utilizing just neural information. Although these neurotechnologies are promising and numerous clinical preliminaries are in the works, numerous technical challenges should be settled (e.g., gadget portability and wearability, algorithm precision, stability, and generalizability) before widespread interpretation into individual assistive devices for loss of motion. Less work has been done to accomplish lower limb motor control with BCI frameworks, albeit some fundamental work has effectively been finished. EEG-BCI control of independent walking has been shown in VR for a patient with spinal cord injury.[116] Additionally, powered exoskeletons have been created and tried for people with spinal cord injury[117] as well as stroke. Interest is being developed in the feasibility of communicating BCI with a wearable exoskeleton (NASAs X1) to interpret lower limb kinetics and kinematics during overground strolling.[118] This might prompt another design of neural interfaces that can straightforwardly control exoskeletons that aid gait retraining in different neurological conditions.

Researches combining BCI frameworks with FES or robotic therapies for the restoration of motor function have been completed in subacute or incomplete spinal cord injury[119] and stroke.[120] The reasoning for consolidating

BCI frameworks with these treatments is to convey high–repetition upper limb movement[121,122] practice while helping dynamic patient commitment to expand treatment impacts[123] and possibly draw in Hebbian learning systems. Benefits of augmenting conventional FES or robotic treatments with BCI incorporate the capacity to convey treatments to people left with minimal motor work, return of the loss of movement control to the patient, and retrain central to peripheral connections. Some researchers additionally have joined BCI treatments with biofeedback to attempt to assist with forming neural reactions and further drive plasticity.[124]

Proof for the viability of BCI-augmented treatment for motor recuperation after SCI is preliminary. One preliminary of BCI–FES selecting 7 individuals with tetraplegia has been accounted for the upper limb in SCI.[125] These creators found that harmless BCI–FES hand restoration gave preferable neurologic recuperation over FES alone in individuals with subacute cervical SCI.[126] Additional pilot information from 1 participant recommends that training with an intracortical BCI–FES as an assistive gadget might continue to work on leftover capacity and hand use without the gadget in chronic SCI, albeit this work should be recreated. Likewise, non–invasive BCI-exoskeleton retraining in SCI might be useful for re-establishing grasp and ambulation, however, randomized preliminaries of viability still need to be finished. Further preliminaries likewise are expected to set up whether BCI-empowered treatment with FES or mechanical exoskeletons can further develop function in SCI across levels of chronicity and levels of injury culmination.

The stroke literature for functional restoration through BCI-augmented treatment is more powerful. Studies have, as a general rule, observed positive outcomes for people with critical loss of motion utilizing both EEG and magnetoencephalography. In a controlled investigation of 32 patients with constant stroke, another gathering showed that BCI control of an orthosis worked on utilitarian additions of the paretic hand comparative with their matched controls. In another review, analysts straightforwardly analyzed the rehabilitative impacts of EEG-BCI treatment to that of robot-helped treatment in an accomplice of 9 patients with moderate-to-serious upper limb hemiparesis from a chronic subcortical stroke. In like manner, another group reported better functional neurologic recuperation after treatment with consolidated EEG–BCI, robotic exoskeleton, and FES. A continuous examination is trying the achievability of a non–invasive, EEG-based BCI-increased restoration utilizing the MAHI-II exoskeleton (Rice University). Primer outcomes propose the capability of this BCI-robot interface to work on both clinical measures of upper extremity function and advanced robotic measures following 12 weeks of preparation.

Sensory restoration

Cutaneous data sources are critical to the natural motor control as they signal state advances, for example, object contact or take-off, and give key data about slip or contact that guide manual interactions.[127] Without tactile signals, the expertise with which we handle and control objects are seriously compromised.[128] In individuals with lower limb amputation, decreased sensory input makes daily activities, for example, step climbing and strolling on lopsided landscape troublesome and hazardous with a prosthetic limb.[129] There is likewise proof that the absence of sensory input adds to phantom limb pain.[130] Regardless of the basic idea of somatosensation in motor control, BCIs have ordinarily just given visual input to the client. Critically, huge continuous work is pointed toward making sensorized prosthetics that can give the data important for stimulation.

Spinal cord stimulation (SCS) is regularly used to treat complex pain by conveying electrical pulses in the epidural space.[131] In a new arrangement of preliminaries,[132] SCS was utilized to target lateral structures in the lumbar and cervical spinal cord,[133] which probably incorporate the dorsal rootlets.[134] These examinations, and those of others, show that SCS with patterned stimulation can prompt naturalistic sensations[135] and lessen phantom limb pain.[136] Primer work has shown that stimulation of the dorsal rootlets in people with upper limb removals can cause an assortment of sensations that are steady after some time, limited to the explicit region of the hand, and are evaluated to such an extent that greater stimulation prompts to stronger sensations. The considerable achievement has been accomplished that helps to empowering BCI control of a robotic arm and empowered subjects with tetraplegia to naturally control the placement of the hand, the direction of the wrist for coordinated functioning of upper extrimities.

Ethical considerations of BCI use in rehabilitation

Various ethical contemplations can possibly affect the utilization of BCI in the recovery setting as either assistive or rehabilitative gadgets.[137] Where the investigational BCI or end effector gadget is invasive, patient independence and informed consent are fundamental.[138] Thorough discussions of dangers and advantages, particularly for surgeries that are not therapeutically essential, can assist with alleviating moral worries. Both the danger of no advantage and the hazard for loss of capacity ought to be thought of and talked about. Civil rights contemplations encompass the utilization of costly innovation in an environment of restricted resources. Cost to society for the turn of events and sending of costly assistive or restorative gadgets should be balanced with the cost to the patient of not creating or giving neurotechnology to diminish disability. Most importantly, the utilization of BCI-empowered assistive gadgets raises the possibility of changing a client's mental self-portrait, identity, and at last the probability of assimilating and embracing everyday utilization of the device. Subsequently, remembering end-users for the design process isn't just a useful issue, e.g., to augment acceptance by including design highlights essential to patients, however, it is additionally an ethical one.

Conclusion

Virtual reality (VR), Augmented reality (AR), as well as Brain–Computer interface technology (BCI), are rapidly observing more extensive acknowledgment in medical services as the medical community is becoming more aware of its advantages as a promising device for different progressive medical care applications, beginning from remote surgeries, clinical treatment, preventive medication, medical education, and training to physical and psychosocial restoration. Virtual, augmented reality interventions just as BCI are turning out to be more available and they are an intriguing and emerging strategy to conceivably improve cognitive function and animate personal memory, as well as to advance memory and potentially the quality of life of an individual. An overall restriction to a few innovative frameworks is that a populace with mental, visual, perceptual impedances might think that they are hard to utilize. Safety issues are likewise basic to consider, particularly given that primary clients have functional deficits that might make them defenseless against mechanical breakdown, particularly if unsupervised. Additional exploration is needed to additionally research robotic and VR frameworks to figure out which plans are generally effective, the proper dosing and timing of intervention, as well as motivation and adherence to restorative programs. Systemically, current research endeavors in the field of advanced robotics and VR are restricted by heterogeneity among interventional approaches, result measures, and control groups, restricting the capacity to pool and compare studies. Also, these investigations for the most part use little sample sizes, and scarcely any examinations have inspected whether impacts are sustained. Additionally, there is frequently restricted financing in the area of rehabilitation research, and treatment utilizing VR, AR robotics, BCI frameworks is at present not well reimbursed. Compounding the difficulties of clinical exploration in robotic recovery is the consistent and fast headway of innovation. alternative pathways for the evaluation of novel robotic frameworks might be thought about so patients can profit from modern innovative advances while guaranteeing patient security.

References

1. Dautenhahn K. Socially intelligent robots: dimensions of human-robot interaction. *Philos Trans R Soc.* 2007;362(1480):679–704.
2. Schultheis MT, Himelstein J, Rizzo AA. Virtual reality and neuropsychology: upgrading the current tools. *J Head Trauma Rehabil.* 2002;17(5):378–394.
3. Burdea GC. Virtual rehabilitation–benefits and challenges. *Methods Inf Med.* 2003;42(5):519–523.
4. Langhorne P, Coupar F, Pollock A. Motor recovery after stroke: a systematic review. *Lancet Neurol.* 2009;8(8):741–754.
5. Katz N, Ring H, Naveh Y, Kizony R, Feintuch U, Weiss PL. Interactive virtual environment training for safe street crossing of right hemisphere stroke patients with unilateral spatial neglect. *Disabil Rehabil.* 2005;27(20):1235–1243.
6. Shih JJ, Krusienski DJ, Wolpaw JR. Brain-computer interfaces in medicine. *Mayo Clin Proc.* 2012;87(3):268–279.
7. Chee Siang A, Sakel M, et al. Use of brain computer interfaces in neurological rehabilitation. *Br J Nurs.* 2011;7(3):523–528. doi:10.12968/bjnn.2011.7.3.523.
8. Stein J, Narendran K, McBean J, Krebs K, Hughes R. Electromyography-controlled exoskeletal upper-limb-powered orthosis for exercise training after stroke. *Am J Phys Med Rehabil.* 2007:255–261.
9. Esquenazi A, Talaty M, Packel A, Saulino M. The ReWalk powered exoskeleton to restore ambulatory function to individuals with thoracic-level motor-complete spinal cord injury. *Am J Phys Med Rehabil.* 2012;91:911–921.
10. Chen Y, Fanchiang HD, Howard A. Effectiveness of virtual reality in children with cerebral palsy: a systematic review and meta-analysis of randomized controlled trials. *Phys Ther.* 2018 Jan 1;98(1):63–77.
11. Parsons TD, Rizzo AA, Rogers S, York P. Virtual reality in paediatric rehabilitation: a review. *Dev Neurorehabil.* 2009;12(4):224–238.
12. Sherman W, Craig A. *Understanding Virtual reality: Interface, application, and Design.* California: Morgan Kaufmann; 2002.
13. Schultheis MT, Rizzo AA. The application of virtual reality technology in rehabilitation. *Rehabil Psychol.* 2001;46:296–311. doi:10.1037/0090-5550.46.3.296.
14. Laver KE, Lange B, George S, Deutsch JE, Saposnik G, Crotty M. Virtual reality for stroke rehabilitation. *Cochrane Database Syst Rev.* 2017 Nov 20;11.
15. Parsons TD. Virtual Reality for Enhanced Ecological Validity and Experimental Control in the Clinical, Affective and Social Neurosciences. *Front Hum Neurosci.* 2015;9:660.
16. Wang Z-R, Wang P, Xing L, Mei lP, Zhao J, Zhang T. Leap motion-based virtual reality training for improving motor functional recovery of upper limbs and neural reorganization in subacute stroke patients. *Neural Regen Res.* 2017;12:1823–1831. doi:10.4103/1673-5374.219043.
17. Orihuela-Espina F, Fernández del Castillo I, Palafox L, Pasaye E, Sánchez-Villavicencio I, Leder R, et al. Neural reorganization accompanying upper limb motor rehabilitation

from stroke with virtual reality-based gesture therapy. *Top Stroke Rehabil.* 2013;20:197–209. doi:10.1310/tsr2003-197.

18. Wright WG, McDevitt J, Tierney R, Haran FJ, Appiah-Kubi KO, Dumont A. Assessing subacute mild traumatic brain injury with a portable virtual reality balance device. *Disabil Rehabil.* 2017 Jul;39(15):1564–1572.

19. Wright WG, Tierney RT, McDevitt JJ. Visual-vestibular processing deficits in mild traumatic brain injury. *Vestib Res.* 2017;27(1):27–37.

20. Stanney KM. *Handbook of Virtual Environments: Design, Implementation, and Applications.* New Jersey: Erlbaum Assoc; 2002.

21. Arthur K, Booth KS, Ware C. Evaluating human performance for Fishtank Virtual Reality. *ACM Transactions on Information Systems.* 1993;11:239–265. doi:10.1145/159161.155359.

22. Cruz-Neira C, Sandin DJ, DeFanti TA, Kenyon RV, Hart JC. The CAVE automatic virtual environment. *Communications.* 1992;38:64–72.

23. Carrozzo M, Lacquaniti F. Virtual reality: a tutorial. *Electroencephalog. Clin Neurophysiol.* 1998 Feb;109(1):1–9.

24. Rizzo AA, Kim G. A SWOT analysis of the field of virtual rehabilitation and therapy. Presence: Teleoperators and Virtual Environments Volume 14 Issue 2 April 2005 pp 119–146. https://doi.org/10.1162/1054746053967094.

25. Rizzo AA, Schultheis MT, Kerns K, Mateer C. Analysis of assets for virtual reality applications in neuropsychology. *Neuropsych Rehab.* 2004;14:207–239. doi:10.1080/09602010343000183.

26. Dahdah MN, Bennett M, Prajapati P, Parsons TD, Sullivan E, Driver S. Application of virtual environments in a multi-disciplinary day neurorehabilitation program to improve executive functioning using the Stroop task. *Neuro Rehabilitation.* 2017;41(4):721–734.

27. Dvorkin AY, Ramaiya M, Larson EB, et al. A "virtually minimal" visuo-haptic training of attention in severe traumatic brain injury. *Neuroeng Rehabil.* 2013 Aug 9;10:92.

28. Cuthbert JP, Staniszewski K, Hays K, Gerber D, Natale A, O'Dell D. Virtual reality-based therapy for the treatment of balance deficits in patients receiving inpatient rehabilitation for traumatic brain injury. *Brain Inj.* 2014;28(2):181–188.

29. Besnard J, Richard P, Banville F, et al. Virtual reality and neuropsychological assessment: the reliability of a virtual kitchen to assess daily-life activities in victims of traumatic brain injury. *Appl Neuropsychol Adult.* 2016;23(3):223–235.

30. Robitaille N, Jackson PL, Hébert LJ, et al. A Virtual Reality avatar interaction (VRai) platform to assess residual executive dysfunction in active military personnel with previous mild traumatic brain injury: proof of concept. *Disabil Rehabil Assist Technol.* 2017 Oct;12(7):758–764.

31. Teel E, Gay M, Johnson B, Slobounov S. Determining sensitivity/specificity of virtual reality-based neuropsychological tool for detecting residual abnormalities following sport-related concussion. *Neuropsychology.* 2016 May;30(4):474–483.

32. Pietrzak E, Pullman S, McGuire A. Using Virtual Reality and Videogames for Traumatic Brain Injury Rehabilitation: a Structured Literature Review. *Games Health J.* 2014 Aug;3(4):202–214.

33. Agten CA, Dennler C, Rosskopf AB, et al. Augmented reality–guided lumbar facet joint injections. *Investig Radiol.* 2018;53(8):495–498.

34. El-Hariri H, Pandey P, Hodgson AJ, et al. Augmented reality visualisation for orthopaedic surgical guidance with pre- and intra-operative multimodal image data fusion. *Healthc Technol Lett.* 2018;5(5):189–193.

35. Kuzhagaliyev T, Clancy NT, Janatka M, et al. *Medical Imaging 2018: Image-guided procedures, Robotic interventions, and Modeling.* Augmented reality needle ablation guidance tool for irreversible electroporation in the pancreas. Houston, TX, USA; 13 March 2018.

36. Solbaiti M, Passera K, Rotillo A. Endo-sight the first AR guided ablative system. doi:10.1186/s41747-018-0054-5. https://www.investhorizon.eu/news/endosight-augmenting-interventional-oncology-202. [Accessed December 2018].

37. Zhao Z, Phyhonen J, et al. Augmented reality technology in image-guided therapy: State-of-the-art review. *J Mech Eng.* 2021. Endosight home page, https://www.endo-sight.it/.
38. Mojica CMM, Garcia JDV, Navkar NV, et al. A prototype holographic augmented reality interface for imageguided prostate cancer interventions. *VCBM* Granada, Spain; 225 December 2018:17–21.
39. Park BJ, Hunt SJ, Nadolski GJ, et al. 3D Augmented reality-assisted CT-Guided interventions: system design and preclinical trial on an abdominal phantom using HoloLens 2, 2020. arXiv preprint arXiv:2005.09146 (2020).
40. Solbiati M, Passera KM, Rotilio A, et al. Augmented reality for interventional oncology: proof-of-concept study of a novel high-end guidance system platform. *Eur Radiol Exp.* 2018;2:18.
41. Khalid MB, Rao NI, Rizwan-i-Haque I, Munir S, Tahir F. Towards a Brain Computer Interface Using Wavelet Transform with Averaged and Time Segmented Adapted Wavelets. *Proceedings of the 2nd International Conference on Computer, Control and Communication (IC4'09) Karachi*; February 2009:1–4.
42. Kubanek J, Miller K, Ojemann J, Wolpaw J, Schalk G. Decoding flexion of individual fingers using electrocorticographic signals in humans. *J Neural Engin.* 2009;6.
43. Arle JE, Shils JL, Malik WQ. *Paper presented at: Engineering in Medicine and Biology Society (EMBC), 2012 Annual International Conference of the IEEE.* Localized stimulation and recording in the spinal cord with microelectrode arrays; 2012.
44. Thakor NV. Translating the brain-machine interface. *Sci Transl Med.* 2013;5:210ps217.
45. Irani F, Platek SM, Bunce S, Ruocco AC, Chute D. Functional near infrared spectroscopy (fNIRS): an emerging neuroimaging technology with important applications for the study of brain disorders. *Clin Neuropsychologist.* 2007;21:9–37.
46. Olson JD, Wander JD, Johnson L, et al. Comparison of subdural and subgaleal recordings of cortical high-gamma activity in humans. *Clin Neurophysiol.* 2016;127:277–284.
47. Olson JD, Wander JD, Darvas F. Demonstration of motor-related beta and high gamma brain signals in subdermal electroencephalography recordings. *Clin Neurophysiol.* 2017;128:395–396.
48. Schalk G, Wolpaw JR, McFarland DJ, Pfurtscheller G. EEG-based communication: presence of an error potential. *Clin Neurophysiol.* 2000;111:2138–2144.
49. McFarland DJ, McCane LM, Wolpaw JR. EEG-based communication and control: short-term role of feedback. *IEEE Trans. Rehabil Eng.* 1998;6:7–11.
50. Widge AS, Moritz CT, Matsuoka Y. *Direct Neural Control of Anatomically Correct Robotic hands. Brain-Computer Interfaces.* Berlin, Heidelberg: Springer-Verlag; 2010:105–119.
51. Fetz EE. Volitional control of neural activity: implications for brain–computer interfaces. *J Physiol.* 2007;579:571–579.
52. Miller KJ, Schalk G, Fetz EE, Den Nijs M, Ojemann JG, Rao RP. Cortical activity during motor execution, motor imagery, and imagery-based online feedback. *Proc Natl Acad Sci USA.* 2010;107:4430–4435.
53. Batula AM, Mark JA, Kim YE, Ayaz H. Comparison of brain activation during motor imagery and motor movement using fNIRS. *Comput Intell Neurosci.* 2017;2017.
54. Sharma G, Friedenberg DA, Annetta N, et al. Using an artificial neural bypass to restore cortical control of rhythmic movements in a human with quadriplegia. *Sci Rep.* 2016;6:33807.
55. Collinger JL, Boninger ML, Bruns TM, Curley K, Wang W, Weber DJ. Functional priorities, assistive technology, and brain-computer interfaces after spinal cord injury. *J Rehabil Res Dev.* 2013;50:145–160.
56. Wodlinger B, Downey JE, Tyler-Kabara EC, Schwartz AB, Boninger ML, Collinger JL. Ten-dimensional anthropomorphic arm control in a human brain-machine interface: difficulties, solutions, and limitations. *J Neural Eng.* 2015;12.

57. Friedenberg DA, Schwemmer MA, Landgraf AJ, et al. Neuroprosthetic-enabled control of graded arm muscle contraction in a paralyzed human. *Sci Rep.* 2017;7:8386.
58. Wu J, Casimo K, Caldwell DJ, Rao RP, Ojemann JG. *Paper presented at: Neural Engineering (NER), 2017 8th International IEEE/EMBS Conference on.* Electrocorticographic dynamics predict visually guided motor imagery of grasp shaping; 2017.
59. Venkatakrishnan A, Francisco GE, Contreras-Vidal JL. Applications of brain–machine interface systems in stroke recovery and rehabilitation. *Curr Phys Med Rehabil Rep.* 2014;2:93–105.
60. Friedenberg DA, Bouton CE, Annetta NV, et al. Big data challenges in decoding cortical activity in a human with quadriplegia to inform a brain computer interface. *Conf Proc IEEE Eng Med Biol Soc.* 2016:3084–3087.
61. Friedenberg DA, Schwemmer M, Skomrock N, et al. Neural decoding algorithm requirements for a take-home brain computer interface. *Conf Proc IEEE Eng Med Biol Soc.* 2018 in press.
62. Downey JE, Weiss JM, Muelling K, et al. Blending of brain-machine interface and vision-guided autonomous robotics improves neuroprosthetic arm performance during grasping. *J Neuroeng Rehabil.* 2016;13:28.
63. Knutson JS, Fu MJ, Sheffler LR, Chae J. Neuromuscular electrical stimulation for motor restoration in hemiplegia. *Phys Med Rehabil Clin North Am.* 2015;26:729.
64. Ragnarsson K. Functional electrical stimulation after spinal cord injury: current use, therapeutic effects and future directions. *Spinal Cord.* 2008;46:255.
65. Peckham PH, Keith MW, Kilgore KL, et al. Efficacy of an implanted neuroprosthesis for restoring hand grasp in tetraplegia: a multicenter study. *Arch Phys Med Rehabil.* 2001;82:1380–1388.
66. Pool D, Elliott C, Bear N, et al. Neuromuscular electrical stimulation-assisted gait increases muscle strength and volume in children with unilateral spastic cerebral palsy. *Dev Med Child Neurol.* 2016;58:492–501.
67. Ajiboye AB, Willett FR, Young DR, et al. Restoration of reaching and grasping movements through brain-controlled muscle stimulation in a person with tetraplegia: a proof-of-concept demonstration. *Lancet.* 2017;389:1821–1830.
68. Müller-Putz GR, Scherer R, Pfurtscheller G, Rupp R. EEG-based neuroprosthesis control: a step towards clinical practice. *Neurosci Lett.* 2005;382:169–174.
69. Mulcahey MJ, Betz RR, Kozin S, Smith BT, Hutchinson D, Lutz C. Implantaton of the Freehand system during initial rehabilitation using minimally invasive techniques. *Spinal Cord.* 2004;42:146–155.
70. Pfurtscheller G, Müller GR, Pfurtscheller J, Gerner HJ, Rupp R. 'Thought'—Control of functional electrical stimulation to restore hand grasp in a patient with tetraplegia. *Neurosci Lett.* 2003;351:33–36.
71. Bouton CE, Shaikhouni A, Annetta NV, et al. Restoring cortical control of functional movement in a human with quadriplegia. *Nature.* 2016;533:247–250.
72. Rohm M, Schneiders M, Müller C, et al. Hybrid brain–computer interfaces and hybrid neuroprostheses for restoration of upper limb functions in individuals with high-level spinal cord injury. *Artif Intell Med.* 2013;59:133–142.
73. Rupp R, Rohm M, Schneiders M, et al. Think2grasp-bci-controlled neuroprosthesis for the upper extremity. *Biomed Tech (Berl.).* 2013. https://doi.org/10.1515/bmt-2013-4440.
74. Grimm F, Walter A, Spüler M, Naros G, Rosenstiel W, Gharabaghi A. Hybrid neuroprosthesis for the upper limb: combining brain-controlled neuromuscular stimulation with a multi-joint arm exoskeleton. *Front Neurosci.* 2016;10:367.
75. Burke D, Gorman E, Stokes D, Lennon O. An evaluation of neuromuscular electrical stimulation in critical care using the ICF framework: a systematic review and meta-analysis. *Clin Respir J.* 2016;10:407–420.

76. Stein C, Fritsch CG, Robinson C, Sbruzzi G, Plentz RDM. Effects of electrical stimulation in spastic muscles after stroke: systematic review and meta-analysis of randomized controlled trials. *Stroke*. 2015;46:2197–2205.
77. Marquez-Chin C, Marquis A, Popovic MR. EEG-triggered functional electrical stimulation therapy for restoring upper limb function in chronic stroke with severe hemiplegia. *Case Rep Neurol Med*. 2016;2016.
78. Rodriguez M, Pierre C, Couve S, et al. Towards brain–robot interfaces in stroke rehabilitation. *PLoS One*. 2011;6:1–17.
79. Takahashi M, Takeda K, Otaka Y, et al. Event related desynchronization-modulated functional electrical stimulation system for stroke rehabilitation: a feasibility study. *J Neuroeng Rehabil*. 2012;9:56.
80. Laver KE, Lange B, George S, Deutsch JE, Saposnik G, Crotty M. Virtual reality for stroke rehabilitation. *Stroke*. 2018 STROKEAHA.117.020275.
81. Knaut LA, Subramanian SK, McFadyen BJ, Bourbonnais D, Levin MF. Kinematics of pointing movements made in a virtual versus a physical 3-dimensional environment in healthy and stroke subjects. *Arch Phys Med Rehabil*. 2009;90:793–802.
82. Collinger JL, Wodlinger B, Downey JE, et al. High-performance neuroprosthetic control by an individual with tetraplegia. *Lancet*. 2013;381:557–564.
83. Tidoni E, Abu-Alqumsan M, Leonardis D, et al. Local and remote cooperation with virtual and robotic agents: a P300 BCI study in healthy and people living with spinal cord injury. *IEEE Trans Neural Syst Rehabil Eng*. 2017;25:1622–1632.
84. Saleh S, Fluet G, Qiu Q, Merians A, Adamovich SV, Tunik E. Neural patterns of reorganization after intensive robot-assisted virtual reality therapy and repetitive task practice in patients with chronic stroke. *Front Neurol*. 2017;8:452.
85. Pfurtscheller G, Guger C, Müller G, Krausz G, Neuper C. Brain oscillations control hand orthosis in a tetraplegic. *Neurosci Lett*. 2000;292:211–214.
86. Lee K, Liu D, Perroud L, Chavarriaga R, Millán JR. A brain-controlled exoskeleton with cascaded event-related desynchronization classifiers. *Robotics Autonomous Systems*. 2017;90:15–23.
87. Sakurada T, Kawase T, Takano K, Komatsu T, Kansaku K. A BMI-based occupational therapy assist suit: asynchronous control by SSVEP. *Front Neurosci*. 2013;7:172.
88. Pfurtscheller G, Allison B, Bauernfeind G, et al. The hybrid BCI. *Front Neurosci*. 2010;4:30.
89. Galán F, Nuttin M, Lew E, et al. A brain-actuated wheelchair: asynchronous and non-invasive brain–computer interfaces for continuous control of robots. *Clin Neurophysiol*. 2008;119:2159–2169.
90. Raspopovic S, Capogrosso M, Petrini FM, et al. Restoring natural sensory feedback in real-time bidirectional hand prostheses. *Sci Transl Med*. 2014;6:222ra219.
91. Flesher SN, Collinger JL, Foldes ST, et al. Intracortical microstimulation of human somatosensory cortex. *Sci Transl Med*. 2016;8:361ra141.
92. Jezernik S, Colombo G, Keller T, Frueh H, Morari M. Robotic orthosis lokomat: a rehabilitation and research tool. *Neuromodulation*. 2003;6:108–115.
93. Daly JJ, Wolpaw JR. Brain–computer interfaces in neurological rehabilitation. *Lancet Neurol*. 2008;7:1032–1043.
94. Dobkin BH. Brain–computer interface technology as a tool to augment plasticity and outcomes for neurological rehabilitation. *J Physiol*. 2007;579:637–642.
95. Bamdad M, Zarshenas H, Auais MA. Application of BCI systems in neurorehabilitation: a scoping review. *Disabil Rehabil*. 2015;10:355–364.
96. Sellers EW, Ryan DB, Hauser CK. Noninvasive brain-computer interface enables communication after brainstem stroke. *Sci Transl Med*. 2014;6:257re257.
97. Wang F, He Y, Qu J, et al. Enhancing clinical communication assessments using an audiovisual BCI for patients with disorders of consciousness. *J Neural Engin*. 2017;14.

98. Hochberg LR, Serruya MD, Friehs GM, et al. Neuronal ensemble control of prosthetic devices by a human with tetraplegia. *Nature.* 2006;442:164–171.
99. Tankus A, Yeshurun Y, Flash T, Fried I. Encoding of speed and direction of movement in the human supplementary motor area. *J Neurosurg.* 2009;110:1304–1316.
100. Wang Y, Hong B, Gao X, Gao S. *Paper presented at: Engineering in Medicine and Biology Society, 2006. EMBS'06. 28th Annual International Conference of the IEEE.* Phase synchrony measurement in motor cortex for classifying single-trial EEG during motor imagery; 2006.
101. Hermes D, Vansteensel MJ, Albers AM, et al. Functional MRI-based identification of brain areas involved in motor imagery for implantable brain–computer interfaces. *J Neural Engin.* 2011;8.
102. Wang W, Collinger JL, Degenhart AD, et al. An electrocorticographic brain interface in an individual with tetraplegia. *PLoS One.* 2013;8:e55344.
103. Klaes C, Kellis S, Aflalo T, et al. Hand shape representations in the human posterior parietal cortex. *J Neurosci.* 2015;35:15466–15476.
104. Broetz D, Braun C, Weber C, Soekadar SR, Caria A, Birbaumer N. Combination of brain-computer interface training and goal-directed physical therapy in chronic stroke: a case report. *Neurorehabil Neural Repair.* 2010;24:674–679.
105. Ramos-Murguialday A, Broetz D, Rea M, et al. Brain–machine interface in chronic stroke rehabilitation: a controlled study. *Ann Neurol.* 2013;74:100–108.
106. Ang KK, Guan C, Phua KS, et al. Brain-computer interface-based robotic end effector system for wrist and hand rehabilitation: results of a three-armed randomized controlled trial for chronic stroke. *Front Neuroeng.* 2014;7:30.
107. Heasman J, Scott T, Kirkup L, Flynn R, Vare V, Gschwind C. Control of a hand grasp neuroprosthesis using an electroencephalogram-triggered switch: demonstration of improvements in performance using wavepacket analysis. *Med Biol Eng Comput.* 2002;40:588–593.
108. Onose G, Grozea C, Anghelescu A, et al. On the feasibility of using motor imagery EEG-based brain–computer interface in chronic tetraplegics for assistive robotic arm control: a clinical test and long-term post-trial follow-up. *Spinal Cord.* 2012;50:599.
109. Kreilinger A, Kaiser V, Rohm M, Rupp R, Müller-Putz GR. BCI and FES training of a spinal cord injured end-user to control a neuroprosthesis. *Biomed Tech (Berl.).* 2013.
110. Downey JE, Brane L, Gaunt RA, Tyler-Kabara EC, Boninger ML, Collinger JL. Motor cortical activity changes during neuroprosthetic-controlled object interaction. *Sci Rep.* 2017;7:16947.
111. Kennedy PR, Bakay RA. Restoration of neural output from a paralyzed patient by a direct brain connection. *Neuroreport.* 1998;9:1707–1711.
112. Spataro R, Chella A, Allison B, et al. Reaching and grasping a glass of water by locked-In ALS patients through a BCI-controlled humanoid robot. *Front Hum Neurosci.* 2017;11:68.
113. Hochberg LR, Bacher D, Jarosiewicz B, et al. Reach and grasp by people with tetraplegia using a neurally controlled robotic arm. *Nature.* 2012;485:372–375.
114. Keith MW, Peckham PH, Thrope GB, Buckett JR, Stroh KC, Menger V. Functional neuromuscular stimulation neuroprostheses for the tetraplegic hand. *Clin Orthop Relat Res.* 1988;233:25–33.
115. Bockbrader M, Kortes MJ, Annetta N, et al. Implanted brain-computer interface controlling a neuroprosthetic for increasing upper limb function in a human with tetraparesis. *PM R.* 2016;8:S242–S243.
116. Wang PT, King CE, Chui LA, Do AH, Nenadic Z. Self-paced brain–computer interface control of ambulation in a virtual reality environment. *J Neural Engin.* 2012;9.
117. Louie DR, Eng JJ, Lam T. Gait speed using powered robotic exoskeletons after spinal cord injury: a systematic review and correlational study. *J Neuroeng Rehabil.* 2015;12:82.

118. He Y, Nathan K, Venkatakrishnan A, et al. An integrated neuro-robotic interface for stroke rehabilitation using the NASA X1 powered lower limb exoskeleton. Paper presented at: Engineering in Medicine and Biology Society (EMBC), 2014 36th Annual International Conference of the IEEE; 2014.
119. Donati AR, Shokur S, Morya E, et al. Long-term training with a brain-machine interface-based gait protocol induces partial neurological recovery in paraplegic patients. Sci Rep. 2016;6:30383.
120. Várkuti B, Guan C, Pan Y, et al. Resting state changes in functional connectivity correlate with movement recovery for BCI and robot-assisted upper-extremity training after stroke. Neurorehabil Neural Repair. 2013;27:53–62.
121. Chen Y-P, Howard AM. Effects of robotic therapy on upper-extremity function in children with cerebral palsy: a systematic review. Dev Neurorehabil. 2016;19:64–71.
122. Dolbow JD, Mehler C, Stevens SL, Hinojosa J. Robotic-assisted gait training therapies for pediatric cerebral palsy: a review. J Rehabil Robotics. 2016;4:14–21.
123. Hu XL, Tong K-y, Song R, Zheng XJ, Leung WW. A comparison between electromyography-driven robot and passive motion device on wrist rehabilitation for chronic stroke. Neurorehabil. Neural Repair. 2009;23:837–846.
124. Young BM, Nigogosyan Z, Walton LM, et al. Changes in functional brain organization and behavioral correlations after rehabilitative therapy using a brain-computer interface. Front Neuroeng. 2014;7:26.
125. Proietti T, Crocher V, Roby-Brami A, Jarrassé N. Upper-limb robotic exoskeletons for neurorehabilitation: a review on control strategies. IEEE Rev Biomed Engin. 2016;9:4–14.
126. Osuagwu BC, Wallace L, Fraser M, Vuckovic A. Rehabilitation of hand in subacute tetraplegic patients based on brain computer interface and functional electrical stimulation: a randomised pilot study. J Neural Engin. 2016;13.
127. Johansson RS, Flanagan JR. Coding and use of tactile signals from the fingertips in object manipulation tasks. Nat Rev Neurosci. 2009;10:345.
128. Monzée J, Lamarre Y, Smith AM. The effects of digital anesthesia on force control using a precision grip. J Neurophysiol. 2003;89:672–683.
129. Johansson R, Hager C, Backstrom L. Somatosensory control of precision grip during unpredictable pulling loads III. Impairments during digital anaesthesia. Exp Brain Res. 1992;89:204–213.
130. Vaso A, Adahan H- M, Gjika A, et al. Peripheral nervous system origin of phantom limb pain. Pain. 2014;155:1384–1391.
131. Cruccu G, Aziz T, Garcia-Larrea L, et al. EFNS guidelines on neurostimulation therapy for neuropathic pain. Eur J Neurol. 2007;14:952–970.
132. Liem L, Russo M, Huygen FJ, et al. A multicenter, prospective trial to assess the safety and performance of the spinal modulation dorsal root ganglion neurostimulator system in the treatment of chronic pain. Neuromodulation. 2013;16:471–482.
133. Deer TR, Grigsby E, Weiner RL, Wilcosky B, Kramer JM. A prospective study of dorsal root ganglion stimulation for the relief of chronic pain. Neuromodulation. 2013;16:67–72.
134. Eldabe S, Burger K, Moser H, et al. Dorsal root ganglion (DRG) stimulation in the treatment of phantom limb pain (PLP). Neuromodulation. 2015;18:610–617.
135. Tan D, Tyler D, Sweet J, Miller J. Intensity modulation: a novel approach to percept control in spinal cord stimulation. Neuromodulation. 2016;19:254–259.
136. Viswanathan A, Phan PC, Burton AW. Use of spinal cord stimulation in the treatment of phantom limb pain: case series and review of the literature. Pain Practice. 2010;10:479–484.
137. Yuste R, Goering S, Agüera y Arcas B, et al. Four ethical priorities for neurotechnologies and AI. Nature. 2017;551:159–163.
138. Klein E. Informed consent in implantable BCI research: identifying risks and exploring meaning. Sci Engin Ethics. 2016;22:1299–1317.

CHAPTER 8

Machine learning, artificial intelligence technologies, and rehabilitation

Introduction

Machines that display artificial intelligence (AI) with the ability to learn from experience and adapt to changing requirements are slowly growing and being deployed in healthcare settings. The technology is being extensively used in wireless devices to gather patients' vital parameters, medical informatics, and other healthcare environments. Machine learning (ML) can be applied to predict the outcome of treatment procedures based on clinical data and patient-reported data (PROMs). They can be applied to analyze the different factors responsible for the success of any treatment procedure. The identified factors that strongly influence the treatment outcome can help to individually adapt to achieve better patient outcomes.[1,2] Presently, they are being employed in rehabilitation of the knee, foot, and hip where professionals can estimate the success of the treatment outcome based on their experience. These technologies are growing at a fast pace and their deployment is being employed in several industrial applications as well as in the healthcare settings. They could be deployed for continuous remote monitoring of patients who could often neglect their vital parameters and health indications due to error in judgement of careless attitude. It allows to provide immediate assistance and not letting it become too late for the patient in which case it may become a severe health condition.

Big Data analysis in coordination with Artificial Intelligence can improve the results of the patient's data which ensures a better and effective treatment outcome. The data-oriented results are used to monitor the patient's healthcare condition efficiently and minimize side effects based on existing data set patterns.[3] These healthcare practices serve as an advancement in treatment quality. The major role of data and information has consistently been vital for the dynamic facility of medical services. With the expanded digitization in medical care, a gigantic measure of data is additionally produced from the healthcare industry.[4] A massive volume of data is accessible today, which

Modern Intervention Tools for Rehabilitation.
DOI: https://doi.org/10.1016/B978-0-323-99124-7.00002-X

holds the guarantee of supporting a wide scope of medical and health-care assignments. The development of analysis techniques and examining, machine learning and artificial intelligence methods along with different opportunities for changing this data into significant and noteworthy bits of knowledge to help decision making, give top-notch patient consideration, react to ongoing circumstances and save more lives on the clinical front; and upgrade the utilization of assets, work on the cycles and benefits to decrease the expenses on the operational and monetary side of the healthcare industry.[3,4] With the adoption of data analytical techniques into the healthcare system, medical care partners can tackle the force of data not just for analysis of recorded data (expressive investigation), but in addition for anticipating future results (perceptive analysis) and for deciding the best activity for the current circumstance (rigid check-up).[4,5] Traditionally, clinical specialists depended on the restricted measure of information accessible to them and their previous experiences for patient treatment. This is of most extreme significance in healthcare, where a solitary choice can mean the contrast between life and disease.[2,6]

Accessibility of data from various sources today offers the freedom to have an all-encompassing comprehension of patient health. Making utilization of cutting-edge innovations over this data likewise empowers access to the right information at the perfect opportunity and ideal spot to convey precise care.[6] Healthcare data – which incorporates static data from patient records, diagnostic pictures, and reports; and dynamic data from monitoring screens or remote patient-monitoring – is generally amorphous in nature. It goes away more former the capacity of conventional analytical tools to deal with such complicated and dynamic data.[6] Through big data analytics and artificial intelligence, the data can be handled to get significant experiences that would be applied as a crucial part in saving lives. Besides the effect of technology and large persistent consideration on the health of the patient, big data analytics and artificial intelligence likewise discover their application in life sciences and clinical analysis. Sub-atomic data analysis opens various measurements for the disclosure of new treatment choices.[6] Prescient logical models can be utilized over data to distinguish hereditary infection markers, plan and foster new medications, and evaluate their adequacy.[7,8] It allows several options to the patient as well as the healthcare service provider based on the outcomes available through these technologies which provide a better understanding and wider perspective in the management of the health diseases and providing scope for an early intervention of the disease which may be beneficial for the patient to manage the diseased condition.

Advanced data analytics and machine learning strategies in this manner empower scientists to design imminent clinical preliminaries based on theories created from the investigation of data.[9] Due to its tremendous potential for working on the nature of healthcare, these emerging methods have drawn in expanding consideration in healthcare analysis and practice. A few analyses have added to the use of big data analytics and artificial intelligence in healthcare; however, the writing generally stays scattered. In request to have a careful comprehension of the possibilities of utilizing these innovations in healthcare, and efficient planning study was performed to get result-oriented output. The efficient planning study is considered proper given the accessibility of generous and different work in this area.[10] Throughout the last decade, AI has been applied to various regions, for example, web indexes, machine interpretation frameworks, and clever individual associates. Simulated intelligence has likewise discovered many uses in the clinical field, alongside the boundless utilization of electronic health records and the fast advancement of life sciences, including neuroscience.[11]

The organization of this outrageous heap of data, which might be both organized and complicated, is the most troublesome undertaking. Executing artificial intelligence (AI) calculations and new combination methods would portend well with such a huge volume of data. For sure, utilizing machine learning (ML) approaches, for example, neural networks and other AI calculations to accomplish automated dynamics would be an immense achievement for healthcare. Machine learning is the investigation of computer programs that learn through deduction and examples instead of being unequivocally customized with calculations and measurable models. The significance of machine learning is prominent for implementation of the smart healthcare practices.[12] Data, the foundation of each model, are the main parts for implementing machine learning in healthcare, and the more pertinent the data, the more exact the gauges. Following the data, a reliable logic should be picked based on the health issue to make more precise estimates in the healthcare delivery.[13]

Step by step, healthcare data are growing at an alarming rate. Data dramatically fill in number utilizing well-being data frameworks and managing patients' records. Immense measures of well-being big data are being created regularly on each day. This opens a discussion regarding how to draw importance from this dramatically developing measure of data. The analysis of such data is significant for removing data, acquiring information, and finding stuffed away instances.[13] Given the current investigations and systems, this part presents a bunch of difficulties that are identified with health big data and health big data analytics. Difficulties have a place with advancements, assortment, capacity, total, investigation, sharing, and perception of health data.[14] In this chapter, we have discussed the efficient planning study about the impression of deployment of machine learning (ML), artificial intelligence (AI) in rehabilitation along with the use of the Internet of Things (IoT) in wearable healthcare devices to support the existing research and discuss the future role of emerging technologies in the development of smart healthcare devices which are going to rule the market in the next decade and future years to come with rapid scale technological advancements, development of Microelectromechanical Systems (MEMs) technology and devices at macro-scale being commercialized.

Machine learning for rehabilitation

Rehabilitation is one of the major steps advised by clinicians to aid in the recovery of patients from any major injury, joint diseases, and musculoskeletal conditions. However, it is not economically viable and clinically feasible to provide long-term access to clinicians and hospitals for such a population especially during a pandemic when the patients are advised to keep their visits to the healthcare centers to the minimum. Hence, patients are advised to undergo rehabilitation exercises in a home-based setup.[15] However, failure to continue regular exercise and motivation in patients to follow appropriate suggested rehabilitative treatment protocol could lead to a long time for recovery and enhanced costs for patients and their families. The main reason for this is the lack of feedback, the procedure to follow standard protocols, and the monitoring of patients in the home environment by their physicians. Here machine learning tools, computer-assisted patient rehabilitation, and information technology could be employed for long-term patient monitoring and improving the feedback inpatient requiring continuous clinical care and rehabilitative treatment for an extended time duration. This would improve decision-making thereby relieving the immense burden on the medical support staff and healthcare facilities at large.[2] It would bring a decisiveness in the actions of the healthcare service providers and build long-term trust between the patient and the doctor which may help to improve their present outlook towards the healthcare service delivery patterns being followed across the centres.

Machine learning (ML) algorithms learn from predicting appropriate success and defining corresponding criteria by repetitive training of the model to gain high precision in estimating successful patient treatment outcomes. They are being employed in medical decision-making and predicting the treatment outcome. These models can be employed in real-life settings to evaluate the patient's rehabilitation outcome post-intervention and training procedures. It would monitor the improvement seen in the patient and the requirement of any adjustments in the planned protocols in consultation with the physicians to report any false results and misclassifications found. The machine learning or ML models have found high success in the treatment of knee, hip, ankle, and foot rehabilitation and allow guidance to evolve improved clinical protocols. Treatment of a large number of elderly, home care patients, and the disabled population is a clinical challenge where progressive research is required continuously.[16] With limitations of resource settings and the availability of trained rehabilitation specialists, the challenge is to improve their overall quality of life and social integration. The use of machine learning tools would allow better targeting of rehabilitation procedures, improved outcome, and achievement of targeted goals for the desired population group. Although the success of the rehabilitative procedures is limited by the level of complexity, varied needs of the individual, and the existence of multiple co-morbidities especially in the elderly population group. The availability of large databases, evolving machine learning tools, and continued research exploration would allow better patient rehabilitation outcomes and overcome challenges that limit treatment outcomes.[17,18] It would allow to provide faster, quality healthcare services at an optimized costs to the patients and their guardians thereby relieving them of economic hardship in accessing such advanced healthcare services within their local environment or at home health delivery for chronic diseases.

Predictive models employing machine learning have also been built for the treatment of stroke patients, those suffering from lower back pain (LBP), or other forms of disabilities. Stroke is among the leading causes of death and disability worldwide. Approximately 20–25 percent of stroke survivors present severe disability, which is associated with increased mortality risk. The use of machine learning methods is gaining immense popularity in biomedical research for patient screening, diagnostic and rehabilitative measures. Machine learning tools have been employed in developing regression models to identify stroke patients who suffer from a high disability and are at risk of death. Further, these models can help in improved decision making and providing clinical care to improve the quality of life of patients suffering from such disabilities. Machine learning (ML) applied to rehabilitation systems could be a potential solution to address telerehabilitation for people with chronic LBP if it allows sufficient accuracy in monitoring adherence performance while providing patient care and guidance.[19]

Machine learning (ML) is a field of Artificial Intelligence that permits electronic devices such as computers, laptops and tablets to learn and improve without being expressly modified. ML calculations can likewise break down a lot of data called Big data through electronic health records for illness anticipation and conclusion. Wearable clinical gadgets are utilized to consistently screen a singular's health status and store it in distributed computing.[20] With regards to a recently distributed review, the expected advantages of modern data analytics and machine learning are discussed in this chapter. The limitation of employing a machine learning approach requires a balanced dataset with independent validation so that the dataset matches the predicted outcome and could be employed in clinical diagnosis. Moreover, a patient–centric and physician-friendly approach is required for achieving the desired results. Machine learning approaches may be employed in exercise therapy with modern tools such as virtual reality and augmented reality to predict human activity recognition. Together with deep learning and convolutional neural network (CNN) smart sensor-based rehabilitation exercise recognition systems may be employed to predict precise movement, best results achieved during the performance of any exercise regimen. Since huge amount of data and information is being generated on regular basis from clinical settings, it is important to distinguish wheat from chaff and to discard information or data which is not beneficial and may occupy unnecessary storage space. With cloud computing and artificial data storage tools it is imperative to discard information which is of no potential usage and retain meaningful information.

Artificial intelligence in rehabilitation

Advancements in technology with the development of Artificial Intelligence (AI) and robotics are rapidly advancing the field of rehabilitation research. Smart wearable devices are being employed in collecting patient data and monitoring the long-term health of the patients. They allow the setting of individualized rehabilitation goals and monitoring the progress of an individual. The assistive robots could help people recuperate from long-term illness and allow them to perform daily activities of living. They can also help to bridge gaps between sensory, motor, and cognitive impairments thereby allowing functional independence and improving the overall health of the individual. The AI can assist in saving the critical battery life of connected devices and prevent the discontinuation of essential life-saving services.[21,22] The wearable devices also allow the collection of physiological, biological, and behavioral data from patients to allow companies such as Apple, Nike, or Google associated with the development of these devices to improve their products and adapt them as per the market requirements to meet the needs of the consumer. The technology is being employed rapidly in fall detection devices for monitoring patient gait, smartwatches for interaction with surrounding devices, and intelligent sensors for rehabilitation. It allows alerting the nursing care staff about any health emergency, fatal falls, or sudden patient requirements. The behavioral patterns and characteristics of an individual can be studied over some time with the help of machine learning and artificial intelligence tools.[23] This allows to develop any prediction model and understanding patient variability which helps to determine disease pattern and enhances the knowledge of the healthcare providers in handling the chronic patients.

Throughout the last decade, Artificial Intelligence has been applied to various regions, for example, web search tools, machine interpretation frameworks, and clever individual colleagues. Artificial Intelligence has additionally discovered many uses in the clinical field, along with the far-reaching utilization of electronic health records (EHRs) and the fast improvement of life sciences, including neuroscience.[11] Artificial intelligence has begun upsetting a few regions in medication, from the plan of proof-based treatment plans to the execution of ongoing logical advancements. Simulated intelligence is seen as an increased or a substitute methodology for healthcare experts. The utilization of AI has been proposed to satisfy information prerequisites in tertiary medical centers.[11] Artificial intelligence has additionally been applied in various fields of perceptive health administrations, like mechanical medical procedures, cardiology, disease treatment, and nervous system science.[24] In future, it would become one of the most important areas on which different tools and techniques would rely upon thereby giving an added advantage in their set-up and design.

Features of AI

Artificial intelligence (AI) is steadily changing clinical practice. With late advancements in digitized data securing, machine learning, and processing framework, AI applications are expanding into regions that were recently thought to be just the area of human specialists.[25] The ability to understand is quite possibly the most intensely discussed topic with regards to the use of artificial intelligence (AI) in healthcare. Even though AI-driven frameworks have been displayed to beat people in specific logical undertakings, the absence of the capacity to understand keeps on starting the analysis. However, the huge volume of collected data that needs to be understood is certainly not a simple mechanical issue, rather it raises a large group of clinical, lawful, moral, ethical and cultural inquiries that require exhaustive investigation and research.[26] Hence, it is important to retain useful information and discard unreliable or incorrect information which could lead to misleading results and may influence the outcome in the long run.

AI for healthcare

Each framework that adjusts Artificial Intelligence enjoys an additional benefit of achieving its assignment inside a short enough said. In healthcare, experts have been taking more time, to sum up, discoveries, yet health frameworks that have Machine Learning calculations are utilized to diminish drug revelation times. For instance, creating drugs utilizing clinical test strategies will take clinicians and experts numerous years and massive expense.[27] Accordingly, the utilization of AI to re-establish portions of the disclosure cycle of medication will be less expensive, speedier, and more secure. In any case, it probably won't be feasible to utilize AI innovation in all the medication disclosure processes. Maybe it helps with stages, for example, the most common way of finding new mixtures that can be potential medications. Additionally, AI can be utilized to recognize the utilization of mixtures put away in the research facility that was recently tried. For model, after the flare-up of Ebola in West Africa, Artificial Intelligence innovation was utilized to check open meds that may be updated to battle the sickness.[28]

Two drugs were found inside at some point, yet comparable analysis led by human intelligence requires a long time in years to decipher. Hence the eventual fate of AI in drug creation is the incorporation of in-memory processing innovation together with AI stages to increment the ability to offer sped-up drug revelation and advancement.[29] Artificial intelligence (AI) programs are applied to strategies like diagnostic techniques, treatment convention improvement, patient observing, drug advancement, customized medication in healthcare, and flare-up expectations in worldwide health, as on account of the current COVID-19 pandemic. Artificial intelligence (AI) by and large applies to computational innovations that imitate instruments helped by human intelligence, like the idea, profound learning, transformation, commitment, and tactile understanding. A few devices can execute a job that ordinarily includes human translation and dynamic multitasking.[12]

Interest and advances in clinical AI applications have flooded lately because of the significantly improved processing force of present–day PCs and the immense measure of computerized data accessible for assortment and use.[30] Man–made intelligence is continuously changing clinical practice. There are a few AI applications in medication that can be utilized in a variety of clinical fields, for example, clinical, diagnostic, rehabilitative, careful, and predictive practices.[31] One more basic space of medication where AI is settling on an effect is clinical dynamic and infection finding. AI innovations can ingest, analyze, and report huge volumes of data across various modalities to distinguish infection and guide clinical choices.[32] Computer-based intelligence applications can manage the immense measure of data created in medication and discover new data that would some way or another stay concealed in the mass of clinical big data.[33] The field of healthcare is advancing at speeding up, and this is joined by a critical expansion in the measure of data and difficulties as far as cost and patient results, so AI applications have been utilized to lessen these difficulties.[34] Artificial Intelligence is extremely valuable in addressing tricky healthcare difficulties and offers various benefits over customary data analytics and clinical dynamic methods. Coming up next are the main instances of the job of artificial intelligence (AI) in healthcare:

Artificial intelligence in medical diagnosis

Artificial intelligence can change clinical diagnostics.[34,35] Pointless routine lab testing increments superfluous monetary expenses. Hence, artificial intelligence applications have been utilized to limit the circle of lab analyses that the patient might require. Computer-based intelligence can distinguish the presence of early illness at the earliest opportunity as it can computerize an enormous part of the manual work and accelerate the determination cycle.[35] This could help in reducing operational costs and routine errors in clinical settings.

AI for clinical workflow

Artificial intelligence is at present being utilized to productively oversee work process and analyze imaging. Artificial intelligence can be utilized to work on the clinical work process, support better clinical bits of knowledge, decrease clinical inconstancy, help in setting concentrate on needs, and limit doctor burnout.[35] Artificial intelligence can assume control throughout the tedious assignment of data input so clinicians can zero in on further developing work use, expanding day by day usefulness, and giving the greatest of care to patients.[2] If technology is adopted by healthcare practitioners it could become a major asset in the existing healthcare settings to improve the quality of patient care and health standards prevalent in the market at present.

AI for anticipating hospital acquired infections

Artificial intelligence can standardize the analysis of contaminations with Infection Prevention and Control (IPC) suggestions, and work with the spread of IPC skills. Computer-based intelligence gives freedom to further develop conclusions through target design acknowledgment. Utilizing AI-driven models, clinicians can screen high hazard patients, foresee which patients are probably going to foster focal line contaminations, and mediate to decrease hazard.[36]

AI for development of the next generation of radiology tools

Artificial intelligence can assist with fostering the up-and-coming age of imaging devices that will give precise data and itemized enough to swap the requirement for tissue tests at times.[37] The up-and-coming age of artificial intelligence is relied upon to be more successful in the healthcare framework and there will be further upgrades in execution. These advancements guarantee to expand precision and decrease the number of routine undertakings that exhaust time and exertion.[38]

Big data in healthcare

Today, we are confronting a circumstance wherein we are overflowed with huge loads of data from each part of our life like social exercises, science, work, health, and so forth as were, we can contrast the current circumstance with a data storm. The mechanical advances have helped us in producing an ever-increasing number of data, even to a level where it has become unmanageable with presently accessible advances. This has prompted the making of the term big data to portray data that is huge and unmanageable. To meet our present and future social requirements, we wanted to foster new techniques to coordinate this data and determine significant data. One such unique social need is healthcare.[39]

Big data is a gigantic measure of data that can do something amazing. It has turned into a subject of extraordinary interest for the beyond twenty years on account of an incredible potential that is concealed in it. Different public and private area ventures produce, store, and dissect big data to further develop the administrations they give. In the healthcare business, different hotspots for big data incorporate clinic records, clinical records of patients, consequences of clinical assessments, and gadgets that are a piece of the web of things.[40]

Biomedical analysis likewise produces a critical piece of big data pertinent to public healthcare. This data requires legitimate administration and analysis to determine significant data. In any case, looking for arrangements by investigating big data rapidly becomes similar to discovering a needle in the bundle.[41] There are different difficulties related to each progression of handling big data which must be outperformed by utilizing top-of-the-line processing answers for big data analysis. That is the reason, to give significant answers for working on general health, healthcare suppliers are needed to be completely outfitted with the suitable foundation to deliberately produce and investigate big data. An effective administration, analysis, and translation of big data can change the game by opening new roads for current healthcare.[42]

Big data analytics is reforming the worldwide healthcare industry. As the world aggregates inconceivable volumes of data and health innovation becomes increasingly more basic to the progression of medication, policymakers and controllers are confronted with extreme difficulties around data security and data confidentiality.[43] The utilization of big data in the healthcare business is developing across the globe, and its probability and advantages are obvious. Big data analytics, predictive analytics, artificial intelligence or calculations, machine learning, and profound learning can outfit an enormous volume of datasets. These datasets can be utilized to further develop analysis, illuminate deterrent medication rehearses and lessen antagonistic impacts of medications and different medicines.[44]

Data reconciliation from different sources makes it hard to catch (as shown in Fig. 8.1) or store the data in a particular characterized model. Due to the heterogeneous organizations and the sheer volume of clinical data, it is difficult to process or examine these utilizing conventional clinical innovation and programming. Big data analysis and refined equipment and programming, for example, calculations or neural organizations are important to dissect and acquire useful bits of knowledge from clinical big data, which can be utilized for expectation and clinical proposal.[45]

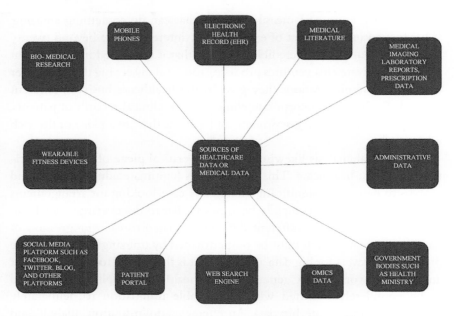

Figure 8.1 Probable sources of big data in the healthcare industry.[31]

Case study: public health system and AI

The dynamism of AI-based innovation in the clinical setting implies that its clients are pushed to adjust to new work processes that coordinate its capacities to emphatically impact health results or, on the other hand, have no certain impact except for rather contorting the treatment pathway. In this manner, regardless of whether an innovation has no demonstrated danger to the patient under given conditions, it should be tried for how it adjusts with client work processes.[45] As per the discussed case study, public healthcare conditions can be improved with the use of AI (Artificial Intelligence).

During the clinical assessment, if a given clinical gadget reacts to the clinical result it expects, there is merit in endeavor human components approval testing considering the climate where it will be utilized. The USFDA suggests that makers decide if the populace utilizing the gadget includes experts or non-professionals, what the clients' schooling levels are, what are the clients are, what practical constraints they might have, and their psychological and tangible conditions.[46] Clinical adequacy for a particular gadget can be fundamentally affected by how the gadget's climate (a controlled lab biological system) is not quite the same as its application climate (an essential health center with restricted web availability). For forefront health laborers with the least computerized proficiency, complex interface capacities on advanced health applications could think twice about the volume of recipients they can react to in a restricted timeframe, hence compromising health results for the local area. Guideline for clinical devices therefore requires express comparable conditions that should be tried and enunciated for its particular ease of use in a general health settings before being put into practice.[46]

Converging machine learning and AI for the future healthcare

The convergence of Big Data Analytics and AI for the future in healthcare depends upon pioneers in healthcare which see gigantic potential in AI and analytics to follow through on the guarantee of greater consideration at a lower cost by engaging their chiefs, business pioneers, clinicians, and medical caretakers by saddling the force of predictive and prescriptive analytics. Numerous healthcare associations are looking to saddle the immense capability of AI to change their clinical and business processes. They look to apply these cutting-edge innovations to sort out a steadily expanding "tidal wave" of organized and unstructured data, and to computerize iterative tasks that recently required manual handling.[47]

As shown in Fig. 8.2, data science steps to perform health data analytics. Healthcare is a living system that generates a significant volume of heterogeneous data. As healthcare systems are pivoting to value-based systems, intelligent and interactive analysis of health data is gaining significance for health system management, especially for resource optimization whilst improving care quality and health outcomes. Health data analytics is being influenced by new concepts and intelligent methods emanating from artificial intelligence and big data.[48] These areas would greatly impact on how the present system works, what possible improvements could be made in their working to make them more efficient and economical with large scale usage and benefits to the stakeholders.[49]

IoT in healthcare

Due to scattered patient data, interconnected networks, higher costs due to system inefficiencies, security, and transparency concerns for the patient the healthcare sector is regulated by strict compliance norms.[49] The Internet of Things (IoT) has rapidly advanced the modern-day healthcare sector thereby allowing continuous and secure monitoring of patients' health and allowing

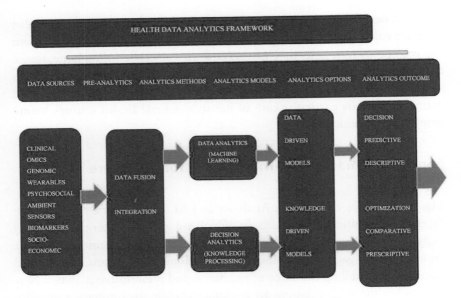

Figure 8.2 Data science steps to perform health data analytics.[48]

the physicians to provide quality healthcare services at affordable costs to the patients and beneficial for all stakeholders involved in the management of well-being.[50,51] Several IoT applications allow enhanced patient–physician interaction with reduced hospital stay duration and reducing the associated costs involved for the patients and their relatives. It mostly supports the critically ill, elderly, and disabled population who require regular monitoring to avoid any healthcare distress and avoid fatality thereby alerting the physician or concerned family members to provide immediate medical care to the patient. The challenges that need to be overcome for the rapid adoption of this technology include data authentication, privacy, and regular secured interaction among the stakeholders.[52,53] With newer technological settings and advances, it is possible that these challenges would be addressed soon and would allow the smooth transition of the IoT infrastructure into the present healthcare devices and systems to make it more efficient, smarter and cost effective.

Conclusion

Intelligent devices have now grown into an important part of modern medicine, twenty years after the publication of the ICF. These have a variety of applications, such as exoskeletons allowing patients with lower-limb paralysis to immediately resume the functions of standing and walking, or prosthetic limbs with sensors allowing patients to feel force feedback. Consequently, disability no longer means suffering. As technology develops and improves, artificial intelligence (AI) and the Internet of Things (IoT) will play an increasingly prominent role in this field, although there is already widespread use in both experimental and clinical applications. Innovation and development in rehabilitation devices mean functional recovery is an achievable goal for an ever-increasing number of people.[54,55]

This chapter additionally came about in better understanding of advancement and approval of models and strategies utilized for healthcare data analysis. Flow research drifts in terms of the use of emerging advances for various regions and measurements of healthcare were analyzed. The discoveries featured healthcare strengths with the most noteworthy exploration interest.[56] Besides, the most broadly applied machine learning calculations and artificial intelligence methods were additionally recognized. Careful assessment of essential investigations prompted distinguishing proof of cutting-edge exploration and disclosure of significant analysis. Suggestions for future exploration incorporate a requirement for more generous contextual analyses and experience papers on the usage of big data analytics and artificial intelligence in healthcare arrangements. This would be logical at the point when healthcare partners and professionals apply these advancements in reality healthcare arrangement, further permitting the revelation of the capability of big data analytics and artificial intelligence for upgrading healthcare quality.[57]

This systematic and evaluation study gives a deprivation of existing checks in the field of big data analytics and artificial intelligence in healthcare. The best test to AI in healthcare isn't whether the advancements will be adequately proficient to be valuable, yet rather guaranteeing their reception day-by-day for the betterment of healthcare practices.

IoT technology is rapidly changing the industry scenario by spreading its wings in all sectors including healthcare. It is being employed as a smart sensing device by the interconnection of devices allowing real-time monitoring of the patient. Certain challenges still exist in its acceptance related to improved storage, higher computational ability, and rapid processing of incoming data. With the integration of rapidly evolving artificial intelligence techniques such as machine and deep learning, the overall system could require human-like reasoning and smart capabilities to enhance decision-making with the help of connected devices. With the advent of cloud computing and data analytics, smart wearable devices are finding rapid market growth and interest. The huge amount of hospital data needs to be accurately mapped within the IoT setup.[57,58] Moreover, sufficient backup and technical know-how are required to overcome large-scale reliance on the technology due to communication network delays, complex functioning which could jeopardize the patient's health condition. If the technology infrastructure is improved rapidly the performance of these integrated technologies would be seen for the treatment of complex diseases and improve service quality. Hence, the future of above technologies with artificial intelligence playing a vital role would definitely help in enhancing and improving the present healthcare settings and make it more viable for the patients and the service providers.

References

1. Mehta N, Pandit A, Shukla S. Transforming healthcare with big data analytics and artificial intelligence: a systematic mapping study. *J Biomed Inform*. 2019;100:103311.
2. Raghupathi W, Raghupathi V. Big data analytics in healthcare: promise and potential. *Heal Inf Sci Syst*. 2014;2:3. https://doi.org/10.1186/2047-2501-2-3.
3. Bates DW, Saria S, Ohno-Machado L, Shah A, Escobar G. Big data in health care: using analytics to identify and manage high-risk and high-cost patients. *Health Aff*. 2014;33:1123–1131. https://doi.org/10.1377/hlthaff.2014.0041.
4. Mohammed EA, Far BH, Naugler C. Applications of the MapReduce programming framework to clinical big data analysis: current landscape and future trends. *BioData Min*. 2014;7:22. https://doi.org/10.1186/1756-0381-7-22.
5. Wang Y, Hajli N. Exploring the path to big data analytics success in healthcare. *J Bus Res*. 2017;70:287–299. https://doi.org/10.1016/j.jbusres.2016.08.002.
6. Costa FF. Big data in biomedicine. *Drug Discov Today*. 2014;19:433–440. https://doi.org/10.1016/j.drudis.2013.10.012.
7. Gutierrez D. *InsideBIGDATA Guide to Healthcare & Life Sciences*. DellEMC and INTEL; 2016. http://en.community.dell.com/cfs-file/__key/telligent-evolutioncomponents-attachments/13-4431-00-00-20-44-33-27/FINAL-White-PaperinsideBIGDATA-Guide-to-Healthcare-and-Life-Sciences.pdf?forcedownload=true.
8. Chen B, Butte AJ. Leveraging big data to transform target selection and drug discovery. *Clin Pharmacol Ther*. 2016;99:285–297. https://doi.org/10.1002/cpt.318.
9. Hudis CA. Big data: are large prospective randomized trials obsolete in the future? *Breast*. 2015;24:S15–S18. https://doi.org/10.1016/j.breast.2015.07.005.
10. Kitchenham B, Charters S. Guidelines for performing Systematic Literature reviews in Software Engineering Version 2.3. *Engineering*. 2007;45:1051. https://doi.org/10.1145/1134285.1134500.
11. Guan J. Artificial intelligence in healthcare and medicine: promises, ethical challenges, and governance. *Chin Med Sci J*. 2019;34(2):76–83.
12. Chattu VK. A review of artificial intelligence, Big Data, and blockchain technology applications in medicine and global health. *Big Data and Cognitive Computing*. 2021;5(3):41.
13. Bani-Salameh H, Al-Qawaqneh M, Taamneh S. Investigating the Adoption of Big Data Management in Healthcare in Jordan. *Data*. 2021;6(2):16.
14. Bollier D, Firestone CM. *The Promise Additionally, Peril of Big Data*. Washington, DC, USA: Aspen Institute, Communications, and Society Program; 2010:1–66.

15. Kankanhalli A, Hahn J, Tan S, Gao G. Big data and analytics in healthcare: introduction to the special section. *Inform Syst Front.* 2016;18:233–235 [Google Scholar].
16. Feldman B, Martin EM, Skotnes T. Big Data in Healthcare Hype and Hope. *Dr Bonnie.* 2012:360. Available online: https://www.ghdonline.org/uploads/big-data-in-healthcare_B_Kaplan_2012.pdf.
17. Frost & Sullivan: Drowning in Big Data? Reducing Information Technology Complexities and Costs for Healthcare Organizations. http://www.emc.com/collateral/analyst-reports/frost-sullivan-reducing-information-technology-complexities-ar.pdf.
18. Knowledge: Big Data and Healthcare Payers. 2013. http://knowledgent.com/mediapage/insights/whitepaper/482.
19. Zenger B: "Can Big Data Solve Healthcare's Big Problems?" HealthByte, February 2012. 2012. http://www.equityhealthcare.com/docstor/EH%20Blog%20on%20Analyticspdf.
20. Benhlima L. Big data management for healthcare systems: architecture, requirements, and implementation. *Adv Bioinformatics.* 2018;2018.
21. Philip Chen CL, Zhang CY. *Inf Sci (NY).* 2014;275:314–347. doi:10.1016/j.ins.2014.01.015.
22. Rathore MM, Ahmad A, Paul A. Te Internet of Tings based medical emergency management using Hadoop ecosystem. Proceedings of the 14th IEEE SENSORS IEEE; November 2015.
23. https://www.optisolbusiness.com/insight/importance-of-big-data-in-healthcare.
24. De Momi E, Ferrigno G. Robotic and artificial intelligence for keyhole neurosurgery: the ROBOCAST project, a multi-modal autonomous path planner. *Proc Inst Mech Eng Part H J Eng Med.* 2010;224(5):715–727.
25. Yu K-H, Beam An L, Kohane Is S. Artificial intelligence in healthcare. *Nat Biomed Eng.* 2018;2(10):719–731.
26. Amann J, et al. Explainability for artificial intelligence in healthcare: a multidisciplinary perspective. *BMC Med Inform Decis Mak.* 2020;20(1):1–9.
27. Tomar D, Agarwal S. A survey on data mining approaches for healthcare. *Int J Bio-Sci Bio-Technol.* 2013;5(5):241–266. doi:10.14257/ijbsbt.2013.5.5.25.
28. Patel S, Patel H. Survey of Data Mining Techniques used in Healthcare Domain. *Int J Inform Sci Tech.* 2016;6(1/2):53–60. doi:10.5121/ijist.2016.6206.
29. Allam S. The impact of artificial intelligence on innovation- an exploratory analysis. *Int J Creative Res Thoughts (IJCRT).* October 2016;4(4):810–814. ISSN:2320-2882. Available at http://www.ijcrt.org/papers/IJCRT1133996.pdf.
30. Dash S, et al. Big data in healthcare: management, analysis, and future prospects. *J Big Data.* 2019;6(1):1–25.
31. Hassan S, et al. Big data and predictive analytics in healthcare in Bangladesh: regulatory challenges. *Heliyon.* 2021:e07179.
32. Carra G, Salluh JIF, da Silva Ramos FJ, Meyfroidt G. Data-driven ICU management: using Big Data and algorithms to improve outcomes. *J Crit Care.* 2020;60:300–304 W.B. Saunders.
33. Price WN, Cohen IG. Privacy in the age of medical big data. *Nat Med.* 2019;25(1).
34. Lee HC, Yoon H-J. Medical Big Data: promises and Challenges. *Kidney Res Clin Pract.* 2017;36(1):3–11.
35. Kaur P, Sharma M, Mittal M. Big data and machine learning-based secure healthcare framework. *Procedia Comput Sci.* 2018;132:1049–1059.
36. Lin YK, Chen H, Brown RA, Li SH, Yang HJ. Healthcare predictive analytics for risk profiling in chronic care: a Bayesian multitask learning approach. *MIS Quarterly.* 2017;41.
37. Archenaa J, Anita EM. A survey of big data analytics in healthcare and government. *Procedia Comput Sci.* 2015;50:408–413.
38. Salomi M, Balamurugan SAA. Need, application and characteristics of big data analytics in healthcare—A survey. *Indian J Sci Technol.* 2016;9(16):1–5.

39. Khanra S, et al. Big data analytics in healthcare: a systematic literature review. *Enterprise Inform Syst.* 2020;14(7):878–912.
40. Tagliaferri SD, Angelova M, Zhao X, Owen PJ, Miller CT, Wilkin T, et al. Artificial intelligence to improve back pain outcomes and lessons learnt from clinical classification approach three systematic reviews. *NPJ Digit Med.* 2020;3(1):1–16.
41. Hamid S The opportunities and risks of artificial intelligence in medicine and healthcare [Internet]. 2016 [cited 2020 May 29]. http://www.cuspe.org/wp-content/uploads/2016/09/Hamid_2016.pdf.
42. Meskò B, Drobni Z, Bényei E, Gergely B, Gyorffy Z. Digital health is a cultural transformation of traditional healthcare. *Mhealth.* 2017;3:38.
43. Shortliffe EH, Sepúlveda MJ. Clinical decision support in the era of artificial intelligence. *JAMA.* 2018;320(21):2199–2200.
44. https://www.scientificworldinfo.com/2021/03/importance-of-artificial-intelligence-in-healthcare.html.
45. Salathé M, Wiegand T, Wenzel M, Kishnamurthy R. *Focus Group on Artificial Intelligence For Health.* ITU, WHO; 2018. https://www.itu.int/en/ITU-T/focusgroups/ai4h/Documents/FG-AI4H_Whitepaper.pdf.
46. Verma A, Rao K, Eluri V. Regulating AI in Public Health: systems Challenges and Perspectives. *ORF Occasional Paper.* 2020;261.
47. https://www.alteryx.com/input/blog/the-convergence-of-artificial-intelligence-ai-and-data-analytics-and-its-implications-for-6.
48. Abidi S, Abidi S. Intelligent health data analytics: a convergence of artificial intelligence and big data. *Healthc Manage Forum.* 2019;32. doi:10.1177/0840470419846134.
49. Bhatia D. *Smart Interfaces For Development of Comprehensive Health Monitoring Systems in Cognitive Computing Systems Applications and Technological Advancements,* 1, Apple Academic Press; 2021:193–217.
50. Bhatia D, Mishra A, Mukherjee M. *Blockchain For 5G-Enabled IoT.* Amalgamation of Blockchain Technology and Internet of Things for Healthcare Applications. Cham: Springer; 2021. https://doi.org/10.1007/978-3-030-67490-8_22.
51. Bhatia D, Bagyaraj S, Karthick AS, Mishra A, Malviya A. *Role of the Internet of Things (IoT) and Deep Learning For the Growth of Healthcare Technology* Trends in Deep Learning, Elsevier; 2020:113–127.
52. Paul S, Bhatia D. *Smart Healthcare For Disease Diagnosis and Prevention.* Elsevier Publisher; 2020 ISBN: 9780128179130.
53. Bhatia D, Paul S. *"Sensor Fusion and Control Techniques For bio-rehabilitation" Bioelectronics and Medical Devices,* 1, Elsevier; 2019:615–632.
54. Qamar S, Abdelrehman AM, Elshafie HEA, Mohiuddin K. Sensor-Based IoT Industrial Healthcare Systems. *Int J Scientific Eng Sci.* 2018;11(2):29–34.
55. Kumar SM, Majumder D. Healthcare Solution based on Machine Learning Applications in IoT and Edge Computing. *Int J Pure Appl Math.* 2018;119(16):1473–1484.
56. Li X, Dunn J, Salins D, et al. Digital health: tracking physiomes and activity using wearable biosensors reveals useful health-related information. *PLoS Biol.* 2017;15(1):e2001402.
57. Steinhubl SR, Feye D, Levine AC, Conkright C, Wegerich SW, Conkright G. Validation of a portable, deployable system for continuous vital sign monitoring using a multiparametric wearable sensor and personalized analytics in an Ebola treatment center. *BMJ Global Health.* 2016;1(1):e000070.
58. Selvaraj N, Narasimhan R. Automated prediction of the apnea-hypopnea index using a wireless patch sensor. Presented at: Engineering in Medicine and Biology Society (EMBC), 2014 36th Annual International Conference of the IEEE. IL, USA (26–30 August 2014).

CHAPTER 9

Exploring the frontiers of assistive technologies in rehabilitation

Background

In medical history, medical advice was given by physicians over the telephone. Traditionally healthcare services have two major drawbacks. Firstly, they are not available all the time and everywhere. Ailing individuals must visit the caregivers or vice versa to start the treatment. This puts a constraint on elderly and/or disabled people living alone and requiring sudden medical attention to thwart possible long-term handicaps. Secondly, the prevailing healthcare infrastructure and personnel are insufficient to cater to the needs of the increasing population. To address the above challenges and to ensure maximum coverage, quality, and efficiency of healthcare services to everyone, everywhere, and all the time, pervasive healthcare solutions have been proposed that use wearable sensors, wireless communications, and mobile computing to achieve their objectives. Smartphones can also play a leading role in various implementations with the help of sensors like accelerometer, proximity sensor, gyroscope, microphone, camera, light sensors, barometer, magnetometer etc. Smartphones can be used to capture users' vital signs and alarm them accordingly. In case of serious injuries or emergencies, caregivers can reach the patient by tracking his GPS. Sensor data collected over a longer period is analysed through intelligent techniques to find out disorders that are not readily observable during the usually short patient–doctor interactions. Smartphones can also do some of the required processing and act as the interface for interactions between machines or things and people.

Nowadays, we all experience and document rehabilitation using digital health technologies. This digital health is defined as the use of a wide range of technologies that may have (but are not limited to) information and communication technology, including the internet, smartphones/mobile health (mHealth), wearable devices, telehealth and telemedicine. For patients, these technologies may improve their ability to connect and access care, providing information to lay people, helping them share their health and

Modern Intervention Tools for Rehabilitation.
DOI: https://doi.org/10.1016/B978-0-323-99124-7.00009-2

illness experiences, and decreasing the financial burden of healthcare. From a therapist's point of view, such technologies can support the collection of clinical information required by rehabilitation experts to guide treatment. Digital health bases itself on implementing and leveraging information and communication technologies (ICTs) to deliver and scale healthcare to the masses. These technologies enable health care delivery to remote patients and proactively support people's wellbeing by avoiding more expensive treatment. The use of digital health tools in healthcare has become ubiquitous and increasingly important. Studies have highlighted their potential to improve: therapeutic interventions and outcomes,[1] self-management,[2] and access to remote consultations.[3] In rehabilitation medicine, digital therapeutics is frequently used for the digital form of care. Digital therapeutics offer new healthcare tools that will impact all stages of the healthcare path, from health maintenance to disease treatment and rehabilitation. Digital therapeutics can contribute to ensuring long-term quality in rehabilitation. Online appointments allow mobile rehabilitation and can be provided over long distances. Controlled by eye movement, a computer can, for example, take over writing and speaking. In some literature, telerehabilitation terminology defines a branch of telemedicine, including information and communication technologies (ICTs). In advanced cases, remote control technologies such as robotics directly provide remote rehabilitation activities. Telerehabilitation includes teleconsultation, telecare at home, telemonitoring, teletherapy, mental telehealth and telelearning. During pandemics and post-pandemic, these measures allow continuity of care for patients who can benefit from remote consultations while ensuring more excellent protection for those who are members of vulnerable groups.

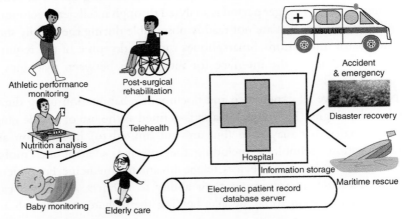

Examples of digital health applications supported by telemedicine.

The following sections define three overlapping categories of digital health care: telemedicine, telehealth, and e-health.

Telemedicine—providing health care services, clinical information, and education over a distance using telecommunication technology—existed long before the internet. The term was coined with the meaning of "healing at a distance", i.e., using Information and Communication Technologies (ICT) to improve patient outcomes by increasing access to care and medical information.[4] Beyond this suggestion, the definition is relatively uninformative since it gives no clue about delivering medicine at a distance. The use of ICT to make the transfer and replace the word "medicine" with the term "transfer of medical data" removes the restriction to patient treatment. A more accurate and informative definition of telemedicine is utilising information and telecommunications technology to transfer medical information for diagnosis, therapy and education. The medical information may include images, live video and audio, video and sound files, patient medical records, and output data from medical devices. The transfer may involve interactive video and audio communication between patients and medical professionals or between those professionals without patient participation.

Telehealth is broad term that encompasses the delivery of health services, education, and information through telecommunication technologies. Telehealth enables the delivery of healthcare services over distances, overcoming geographic obstacles and access to care. It does this by utilising a variety of communication channels, including videoconferencing, phone calls, emails, mobile applications, and secure messaging. Telehealth extends beyond clinical care and includes services focused on promoting and protecting health.

Telerehabilitation (TR) is a relatively new and developing field of telehealth. TR is the use of telecommunication technology to deliver and support rehabilitation services.[5] It is the clinical application of consultative, preventative, diagnostic, and therapeutic services via two-way or multi-point interactive telecommunication technology. These services include but are not limited to assessment, monitoring, intervention, supervision, education, consultation, and counselling.[6] There are two main components of TR services: rehabilitation service (clinical application) and telecommunication/information technology. Research reported many specific advantages and disadvantages of TR. Advantages of TR include practitioners and patients no longer having to be in the same room, and the distance barrier can be minimised to deliver rehabilitation interventions remotely across the continuum of care. Also, decreased travel to specialised urban health centres, enhanced clinical support in local communities, improved access to specialised services and educational benefits, reduced feelings of isolation for rural clinicians, and improved service stability and multimedia communication. Disadvantages of TR include difficulties encountered by the user as well as with the equipment. User problems may be classified as cognitive, motor, and social. In terms of cognitive factors, technological and communication devices may be complicated for an individual with a neurological disability to operate. For example, many brain injuries involve disturbances in an individual's ability to plan, organise, initiate activities, control impulses, concentrate, problem solve, and recall information.[7]

Telecare is often used to describe the application of telemedicine to deliver medical services to patients in their homes or supervised institutions. Telecare is helpful for community care, assisting disabled people, taking care of children and the elderly, and supporting vulnerable individuals. Very simply, people can enjoy the freedom of being left alone while knowing that assistance is always there if or when needed. Telecare can be as simple as an alarm to call for emergency assistance, or a sophisticated system that monitors the user's health condition, assistive network of devices for various routine tasks, an automated personal assistant that reminds the users of different things such as taking prescribed medication and switching off the gas stove after cooking; and the list of what telecare can do goes on.

In 1999, e-health surfaced as a popular term for Internet-based health care delivery.[8] E-health refers to all forms of electronic health care delivered over the internet, ranging from informational, educational, and commercial "products" to direct services offered by professionals, non-professionals, businesses, or consumers. E-health services encompass the "five Cs": content, connectivity, commerce, community, and clinical care,[9] to which some would add a sixth: computer applications in the form of application service providers, E-health draws on the unique capabilities of the internet while enabling delivery of the clinical services that have characterised telehealth and telemedicine. As a result, e-health is making health care more efficient, allowing patients and professionals to do the previously impossible through the efficiencies of the internet.

The most recent definition for telemedicine, telehealth, and related terms, which came out in 2020 from the US Centres for Medicare & Medicaid Services (CMS), is as follows: "the exchange of medical information from one site to another through electronic communication to improve a patient's health."[10] It provides several benefits, the most important of which include simplified access to health facilities and a reduction in the distance between patient and doctor, especially in geographical areas where the medical services are difficult to reach or in the case of seafarers, who are remote individuals. Moreover, telemedicine may improve access to physicians for patients with mobility problems, such as patients with disabilities, fragile patients, or older patients,[11] and could ideally promote equity of access to health care and quick patient engagement at reduced cost.[12] Indeed, in this particular situation, telemedicine services have proven indispensable in facing the emerging needs of health care in this specific context.[13] Many lives have been saved during the COVID-19 pandemic through the use of telemedicine services, making it possible to avoid physical or face-to-face contact with medical staff, healthcare personnel or other health professionals and by possibly reducing the virus spread and preventing or minimising the risks of contagion either for patients or healthcare personnel.[14,15]

In a face-to-face consultation, a clinician might use some combination of all five senses–sight, sound, touch, smell, and taste–to assess a patient's condition. The first three responses may be used in telerehabilitation through sensory data that is transmitted directly from the patient to the observer. These sensory data are first converted into electrical impulses for transmission to the remote observer. The information (useful data) derived from these senses can be divided into four types:

- text and data;
- audio;
- still (single) images;
- video (sequential images).

Benefits and limitations of telerehabilitation

Telerehabilitation's origins, development, and drivers have many benefits and limitations. We can summarise the principal benefits claimed for telerehabilitation as follows:

- Better access to healthcare;
- Access to better healthcare;
- Improved communication between carers;
- Easier and better-continuing education;
- Better access to information;
- Better resource utilisation;
- Reduced costs.

Better access to healthcare

Extending rehabilitation access to rural communities and disadvantaged populations, poorly served or without these facilities, is still one of the significant drivers of telerehabilitation. Greater convenience to patients by reducing travel and disruption is a benefit sought for and claimed by most projects. Time savings for both patient and carer and faster access to care are similarly easy to demonstrate where they occur. Telecare offers many examples of these benefits.

Access to better healthcare

Any healthcare is obviously better where none existed before, but under this heading, we are looking for improvements in the quality of care. A clear benefit of telerehab is the remote access that patients and their therapists have to specialist advice when it is unavailable locally. Early intervention, more seamless care (including care protocols) and better progress monitoring are additional advantages of telemedicine links involving a primary care clinician, a hospital specialist and a community care nurse.

Improved communication between curers

The shift to digital information offers numerous benefits for carers and their patients. Digitised data such as a patient's previous history, X-rays, test results and notes for the current episode are readily transmitted electronically using standard protocols and technologies such as email. Discharge letters are similarly available without delay. Digital communication provides health-care information that is more accurate, more complete and more timely-attributes of quality that lead to better access and better healthcare.

Easier and better continuing education

Through technology, the provision of healthcare courses, perhaps with awards, for the general public. Countries are now promoting a subsidised scheme for low-income families to help them gain home access to the internet. Low-income groups are often at the most significant risk of disease due to socioeconomic conditions and lifestyle. The internet could be used for health promotion with websites targeting both children and parents. It could also be used to advertise health programmes and provide incentives to encourage take-up. The opportunities are endless.

Better access to information

Better access to information is concerned more with the individual endeav-ouring to "pull" information from the internet and other sources to answer specific questions, like an electronic library, website etc.

Better resource utilisation

Better access to healthcare and access to better healthcare are one side of the access coin. Better resource utilisation is the other side. It is uneconomic to replicate resources in several centres when they have infrequent use. A preferred approach is, therefore, to set up a smaller number of resource sites and make these available to potential users via telemedical links.

Reduced costs

This is the most contentious benefit of telecare that has shown cost-effectiveness. One of the reasons is that these techniques often involve few presenting patients, and it is not clear how costs and benefits scale. There is also evidence of economic benefits from telecare in home healthcare and the care of individuals from remote areas.

Some reported limitations of telerehabilitation include the following:
- Poor relationships between healthcare professionals;
- Impersonal technology;
- Organisational disruption;
- Additional training needs;
- Difficult protocol development.

Poor relationships between healthcare professionals

Telemedicine can represent a threat to status and preferred practices. The likelihood of such threats is enhanced if one of more of the clinical participants is over-enthusiastic and tries to coerce unconvinced colleagues into using the link without due discussion or preparation. Most of the identified errors fall into the "technology" or "bureaucracy driven" category, leaving insufficient emphasis on the clinical benefits. Because of consistent dependency on technology also impacts poor patient-carer relationships. This relation often fails to establish in the case of technophobic patients or professionals. Their incidence is, therefore, most significant in elderly patients whose lack of confidence fuels their confusion.

Organisational disruption

Introducing new technologies and working methods always lead to some disruption and concern about the short-and long-term consequences, for example, fear of increasing the workload, lack of skills and the need to acquire them, and lack of agreed standards.

Additional training needs

Consistent education and training in technology are essential as the system develops and new staff are brought on board. The training usually covers the setting up and use of the equipment, the teleconsultation process, and the production of appropriate documentation for these tasks and for recording the consultation procedures and outcomes.

Difficult protocol development

Protocol or pathway development is one of the most critical and time-consuming aspects of introducing technology in rehabilitation. The effective intervention protocol comes from the holistic and integrated view of the care that arises from a multidisciplinary team.

Mobile health technology

Mobile phone is a source of tremendous growth worldwide and is considered a sub-segment of eHealth. More than 70 percent of the world's cell phone users now reside in developing countries due to their affordability and usability. Many new technologies based on cell phones, also known as mHealth open the doors for novel mHealth systems for remote diagnostics, location-aware services and point-of-care systems, body sensor networks, and remote patient monitoring. Overall mHealth applications promise great potential in improving health care and rehabilitation in developing countries. In general, mHealth is categorised into the following areas:

a) Education and awareness, which uses mobile phones to send health alerts to subscribers to make them more aware of disease symptoms and treatment options relevant to their health history and geographic area. Due to the social stigma attached to the disease, using mobile phones allows a private setting for interested parties to gain more awareness about the disease. mHealth helpline projects allow users to call a helpline number to get medical counselling and advice.

b) Diagnostic support, treatment support, and training: Applications in this domain focus on providing health care workers training in diagnosing patients and to support medical experts and hospitals in their activities. Some mobile applications in this area make use of decision trees to ascertain the exact disease and treatment. Several smartphone-based application equips health workers with mobile hand-held devices, allowing them to communicate with doctors, e.g., in order to obtain medication or to obtain help in diagnosing patients.

c) Disease surveillance and outbreak tracking: These mHealth applications rely on mobile phones' ability to transmit data quickly and reliably with high efficiency. Data concerning diseases like TB, malaria, etc., can help health organisations better track the regions where these diseases have emerged and target medical resources to these geographic areas more effectively.

d) Chronic disease management and treatment compliance applications: Some diseases require a strict medical regimen to be followed to avoid complications for individual patients. Health care workers can monitor patients and immediately identify symptoms in patients that require urgent attention.

e) Remote Data Collection: This area aims to link hospitals and healthcare workers with data collectors and databases that house medical information. This is especially useful for government policymakers who can use the data to decide where to assign medical resources and identify which areas need more attention.

In current scenario, mHealth incorporate the wearable sensors interfacing to mobile phones to sense patient conditions. Body sensor networks (BSNs) collect physiological data from patients and perform distributed and collaborative processing. Systems developed today can be used in rehabilitation, sports medicine, geriatric care, gait analysis, balance evaluation, and sports training. Another solution in mHealth is for remote health care and home monitoring application for the elderly, where relatives need to keep tabs on their health and constantly monitor if they are following their medical regime. In addition, technology is intended to strengthen patients' long-term involvement in the healthcare reporting process.

Challenges in the deployment of mobile health technology

Though this technology has been and is being adopted at a large scale by clinician and rehabilitation experts in the developed world, there are major challenges that prevent their widespread adoption in the developing world. mHealth can relieve the resource burden on more traditional healthcare services providers like local hospitals and clinics and can play an important role at regional, community, and individual levels. There are however many shortcomings and associated challenges in the successful deployment and usage of these technologies at a wide scale in developing countries. The highlight of these challenges:

Technological challenges

There are logistical, political, cultural, and financial barriers when trying to implement technology improvements targeted at the mHealth sector. There are either a severe lack or nonexistent hospital legacy systems in developing nations, which could be used to host mHealth services such as patient records for retrieval over mobile phones, etc. Unlike in the developed world, where medical records are increasingly electronic, medical records are still paper based in the developing world. In the developing world, the proliferation of expensive smartphones is also not that dominant for the general population. It is important for governments and policymakers to invest substantially in training health care workers in the use of new systems, to educate them on their benefits, and to either upgrade or overhaul outdated legacy systems and adapt them to the new technological needs of mHealth initiatives. It is necessary for governments to cautiously invest in these technologies, which will form the backbone of future mHealth applications. In cases where people cannot afford handsets, free handsets and loaned mobiles need to be provided.

Socioeconomic challenges

There are very steep socioeconomic gaps across families and individuals in the developing world. It is necessary to seek maximum participation across all social classes to maximise the effectiveness of mHealth applications. Sometimes, cultural beliefs keep people in under-served communities from actively participating in sharing their health information for fear of being ostracised in their community. It is necessary to start mHealth education initiatives to educate people on the full benefits of Mobile Health technology. For people who cannot afford mobile phones, provisions need to be made to loan their phones to allow them to improve their lifestyle and realise the benefits of such solutions.

Infrastructure challenges

There is a need to continue the expansion of wireless Internet and cellular data infrastructure to keep up with the rapid increase of cell phone users in previously inaccessible terrains. To implement an mHealth environment, it is essential to form the necessary foundation for the inexpensive and inefficient exchange of data, even in the remotest of places.

Security and privacy challenges

Medical records are confidential data, and it is very important to protect them. With the growth of mHealth applications, many valuable applications, systems, and tools have exploded. As with personal computers in the early days, a lot of isolated high-potential tools are emerging. However, since mHealth is still in its infancy, these tools need additional integration and security in order to protect confidential medical records. There are security risks beyond conventional networks when using Mobile Health applications because these devices are prone to being misplaced or stolen. System designers need to enforce security features on the phone, including biometric, fingerprint, or password authentication when a Mobile Health session is initiated. A mobile phone's operating system (OS) needs to be engineered to be resistant to tampering. There is a need in developing countries to adopt a layered security hierarchy in consultation with hospital organisations.

Analytical challenges

It is necessary to provide real-time analysis and evaluation of a patient's health statistics and physiological data logged from remote mobile devices. There has been a lot of research on how to accomplish this. An algorithm for real-time analysis of physiological data on a mobile device allows medical staff to monitor patients' data without being physically near them. In the developing world, the expertise to analyse such data is found lacking. Existing medical staff is not trained to perform analysis of such data. One of the major challenges in developing countries is that up-to-date health trend data is not maintained and is not in electronic form. Staff needs to be trained to analyse these records, provide suitable and real-time feedback when required, and validate the effectiveness of their response.

Language and literacy challenges

In most parts of the developing world, the people most in need of medical support are also illiterate. These applications need to communicate with their users in a language that can be easily understood in a local language. This challenge also needs to be addressed if mHealth applications are to be successfully deployed. For people with poor reading skills, applications that can use alternatives, such as voice messaging for such a target population, can be a possible solution. Companies providing mHealth technology need to invest significantly in research so that all sections of society can benefit from these mHealth innovations. In terms of local language support, translation processes need to be followed to maximise the quality of the information in a language appropriate to a targeted population.

Structural challenges

In developing nations, there is a lack of players with the required expertise to undertake mHealth initiatives since existing organisations do not have a track record of implementing these kinds of applications. The mHealth revolution is very dynamic in nature and it is important for all providers, patients, payers, and health planners to comprehend the change in this model, which would allow access to resources at any time and place, including at a point of care. Patients need to learn how to use mobile devices to maximise health and wellness benefits by being engaged and better informed about mobile applications. All stakeholders in the health field need to work on standards, policies, and functionalities to implement all the necessary structural changes and adjustments to the existing healthcare infrastructure.

Quality of health care challenges

In current platforms supporting mHealth applications, services are delivered best-effort since there are no guarantees on the delivered Quality-of-Service (QoS). Allowing healthcare professionals to support any place, anytime monitoring vital signs and providing real-time feedback poses questions on how the quality of relayed information can be improved. For that, context information is related to a serviced user's actual situation (e.g., location, time, current health status) to improve the delivered QoS. Other initiatives like educating healthcare personnel on how to maintain and infer such information to enhance given QoS must be addressed at structural and all levels.

Legal challenges

Health records are confidential information, and various legal issues need to be addressed to sanitise their handling and dissemination in a Mobile Health scenario. Legislation is not as streamlined and no such accountability act available in the developing world. Even in the developed world, legislation to bring change to health services has taken many years, and is a continually ongoing process. This in itself poses a challenge, as there may be a lengthy legislation process to ensure the confidentiality of patient records.

Smart homes assistive technologies

As people age, the risk of illness and disease increases, and the need for health care and support increases. These challenges are standing at our front door and need to be dealt with in a way that prepares society for future elderly generations. The transformation of buildings into sm art homes,[16] assistive environments for senior citizens,[17] or robotic environments[18] are currently in the focus of the research and development of both academia and industry. Technology can enhance security and autonomy and allow the possibility to stay longer at home. In addition, there is also the possibility for an enhancement in the quality of living and QoL. The current problematic issues of urban elderly home living are challenging, and the involvement of ICT solutions and technologies in the actual living environment of urban societies make themselves safer, healthier and happier. The smart home is an ordinary home that integrates comfort, healthcare, safety, security and energy conservation. In a smart home, it is common to use a remote backend system to monitor users' health data, control home devices, and even support remotely from a hospital or housekeeping companies. For instance, the home living mobility elderly control the light, curtains or air conditioner by simple touch the device interface. Features and services in a smart home are divided into three parts: comfort, health care and security. It is also feasible for the user to check his own physical and activity data through phone or computer. After the user and device authentication, the security services ensure the accuracy of the user's data and protect the user's information. The technology in these homes can be pre-programmed to contribute towards household tasks, controlling windows and lighting, multimedia, and advanced sensors both in flooring and infrared in roofing. Not only does the smart home support older adults still living at home, but this new health service model can also provide better information communicated between the health personnel, services and institutions. For the elderly population, smart homes help monitor physiological parameters noninvasively i.e., without attaching sensors and transducers to the body, which is preferable, especially in emergencies. This facility incorporates a range of features related to the approach to the house, orientation and movement within the house, and building management.

Nowadays, many home appliances are controlled by remote controllers. Voice-operated is an alternative for a person with disabilities and/or the elderly. However, recognising command in a noisy environment becomes difficult and unstable, and expressing some spatial positions with voice command is not easy. A home network system was installed as part of the smart house control and communication network to control lighting, curtains, and windows. Furthermore, the security system has a video phone connected to the front door. Issues of privacy and ethics are also critical in this system. The elderly were uncomfortable with continuous monitoring, which seemed to rob them of their privacy. The main disadvantage of this system is that it must be installed when a house is being built, and the installation is expensive and time-consuming. For emergency cases, such as sudden illness, we will sometimes need to install this type of system in an existing house; therefore, a monitoring system should be easy to install and remove.

Developing a reliable home monitoring system has drawn much attention recently due to the growing demands for integrated health care. Existing approaches to current home monitoring systems often include custom-fit environmental, physiological and vision sensors.[19] Such systems can enable several applications to increase personal safety for elderly patients and facilitate clinicians to diagnose and monitor patients. This new patient-clinician interactive mode improves the reliability and effectiveness of diagnosis, significantly shortens the travel time and hospital stay for patients and reduces the workload for clinicians. The following are essential features of a smart home:

Smart light: These lights can lower the electricity bill. These smart lights can be controlled via smartphone and can set schedules for when these lights turn on and off.

Smart speakers can control speech and tell you the news and weather, read your recipes, even order a pizza and play music. Amazon Echo was the first smart speaker, producing good enough audio to be the primary speaker for an apartment or living room.

Smart plug or switch is the easiest and cheapest way to make any appliance in your home "smart". Connect a lamp to the smart switch and be able to control the lamp from your phone. You can also set schedules for the plugs to turn on and off and link them to other smart home devices.

Home security cameras keep an eye on your house and keep tabs on who's coming and going, all from your phone, tablet or laptop. These cameras will also have night vision and can be linked with other smart home devices. That means the camera will start recording when you leave and stop when you return.

Smart locks will also let you issue temporary passes to others and see who's using your door. Integrate the lock with smart home devices, and you could have your lights turn on automatically when you unlock your door when returning home in the evening. A face recognition system is installed at your door and connected to the internet. Imagine that you are busy with your work at the office and your parents come to your home. And keys are with you, but there will be no problem as you have the Face Recognition System. When someone is at your doorstep, it will capture a small video and send it to your mobile application. If you want to open the door, you can open it while sitting in the office.

Smart thermostat remotely sets the temperature in your house using your smartphone. When it is linked to other smart home devices, such as motion sensors and lights, it will also save your money in the long run by reducing your heating and cooling costs, as you won't be using energy when you're not at home.

Smart video doorbells send a live feed to your smartphone or tablet when someone pushes the button at your front door.

Smart smoke detector can alert your smartphone whenever there's smoke or fire in your house.

Assistive technology for cognition

Assistive technology (AT) helps improve care for people with dementia. It helps informal caregivers to persons with dementia and persons with dementia become more independent, provide autonomy, enhance their safety, and improve their quality of life. In line with this suggestion,[20] suggested that AT would assist the informal caregiver in becoming more supportive and ensure that the patient with dementia remains within the community. Sri et al.[21] defined assistive technology as tools, devices, and systems that a person can use to improve and maintain their functional capabilities and independence, thus helping them address their physical, communication, and cognitive difficulties. These assistive technologies include reminders, domestic systems, automatic lights, and alarm systems.

There are several types of assistive technologies, including pervasive telecare and surveillance systems. Pervasive telecare technologies encompass motion detectors, pressure sensors, and temperature and inactivity-detecting sensors. These sensors automatically relay the signal to a caregiver or the monitoring center, thus enabling real-time access to assistance. Surveillance technologies ensure that the patient is monitored constantly using the Global Positioning Systems (GPS) or electronic tracking chips. The technology alerts the caregiver of the position of the patient. Sri et al.[21] explained that the alarm systems could help locate people with dementia whenever they leave home. Other assistive technologies, such as touchscreen devices, include entertainment features such as music and apps, thus improving their quality of life. More devices have emerged in the market in the recent past, which create a fluid environment for dementia patients, thus supporting their convenient living.[20] Several studies have been conducted investigating how AT helps people with dementia. These studies suggest that caregivers could use AT to look after a person with dementia.

Standing and smart wheelchair

The world today makes a huge difference to people with restricted mobility. Advanced assistive technology in the form of a standing wheelchair and smart wheelchair can be used by a person who relies on a wheelchair for mobility. Prolonged use of a basic wheelchair can lead to significant health issues, like reduced bone density, joint muscle stiffness, circulatory system problems and pressure sores.[22] For these populations, adjustable standing electric/manual wheelchair is essential to reduce the negative effects of prolonged sitting in a wheelchair. Form psychological point of view A standing frame provides alternative positioning to sitting in a wheelchair by supporting the person in the upright position. Because they can help patients do activities in standing positions conveniently such as communicating with staff on counter, withdrawing cash from an ATM and supporting elderly getting up from sitting in wheelchairs comfortably. A manually operated standing wheelchair consists of hand driven mechanism that allows users to drive the wheels of the wheelchair while seated, standing or in the range of position in between. A powered standing wheelchair, consists of an electric motor-driven system controlled by the remote-control system to allow for forward, reverse and sideways propulsion of the wheelchair.

To accommodate the population with low vision, visual field neglect, spasticity, tremor or cognitive deficit, several researchers have used technologies to develop smart wheelchair. A smart wheelchair is a sensor-based powered wheelchair whose motion can be controlled by the users' body parts movement, and sometimes it is navigated by speech and vision. People with cognitive/motor/sensory impairment, whether it is due to disability or disease, rely on power wheelchairs (PW) for their mobility goal. A lack of independent mobility at any age places additional obstacles to pursuing vocational and educational needs. Other keywords used for PW are smart wheelchair, intelligent wheelchair, autonomous wheelchair, and robotic wheelchair. Since some people with disabilities cannot use a traditional joystick to navigate their PW they use alternative control systems like head joysticks, chin joysticks, sip-n-puff, and thought control.[23,24] The benefits of powered wheelchair use have been documented: allowing older adults the freedom to engage with their environment and also relieving burden on caregivers.[3] A smart wheelchair (SW) typically a standard PW base to which a computer and a collection of sensors have been added or a mobile robot base to which a seat has been attached. What makes a wheelchair smart is not just a collection of hardware but specialised computer algorithms that provide the artificial intelligence needed to make split-second decisions about where the wheelchair is heading and what might be in its way. Some examples of machine learning in SWs include the use of neural networks to detect obstacles and reproduce pre-taught routes.

Technological assistance services

Technological assistance refers to the use of equipment or devices (walking sticks, canes, bathroom rails, raised toilet seats, hearing aid, etc.) to allow performance of daily activities in an around the home. This Equipment has the potential to promote independence and improve the quality of life for older adults with ADL/IADL disabilities.[25] This technology can benefit as an assistive and supportive tool for both caretakers and the elderly. Technological assistance is sometimes referred to in various literatures as assistive devices or assistive technology.[26,27] Assistive Technologies (AT) includes "any item, piece of equipment, or product system, whether acquired commercially, modified or customised, that is used to increase, maintain, or improve functional capabilities of individuals with disabilities."[28] The use of devices increases with age, almost doubling each decade after the age of 65 years.[29] These assistive technologies range from common devices such as canes and ramps to more high-tech devices such as electric wheelchairs and devices to monitor and prompt people with disabilities to take medicine and perform other tasks like eating. These technologies are used to compensate for physical or cognitive impairments to enable activities of daily living and to reduce isolation.[30] When it is possible to do so, many people prefer to use assistive devices to complete activities of daily living rather than receive help in performing those tasks.[31] Research suggests that the use of assistive technology devices by older adults can result in a higher perceived quality of life, greater independence, improved functional capacities of daily living and a heightened level of social activity.[32,33]

Gitlin[34] identified three main areas of an older adult's life where assistive technology may be effective. These include (1) before the onset of a disability, playing more of a preventive role, (2) after first experiencing the onset of a disability, and (3) during long term care. Unfortunately, most older adults are not aware of the assistive technology devices available that can dramatically improve their quality of daily living.[32] However, when awareness is gained, many older adults have a tendency to discard the idea of using assistive technology to improve their lives. Whether the reason is a lack of knowledge or a fear of the unknown, some older adults find the use of assistive technology difficult and intimidating. Even after purchases of assistive technology have been made, many older adults may feel a sense of discomfort, frustration or embarrassment. This may be due to unfamiliarity, inadequate training in the use of the product, and feelings of embarrassment due to assistive devices being used in public. In addition, older adults may also not have received information on where to purchase replacement parts or other assistive devices when the need arises.[32,34]

Importance of technological and personal assistance

During 20th century there was a dramatic growth of older adults with disabilities using assistive technologies.[35,36] During the 1980s, those using any equipment rose from 3.3 to 4.1 million and the number relying *only* on equipment more than doubled[35]. By the mid- 1990s, the majority of those with activity of daily living (ADL) limitations used some form of technology, and almost a third (31 percent) used only AT to accommodate their needs.[37] Clinical literature mainly focused on the efficacy of specific devices and their efficacy depends on design, provision of appropriate training, and effective home based assessment.[38–41]

Assistive technology can help individuals with disabilities compensate for lost functions, increase their independence, alleviate pressures on the exiting long- term care system, and thus ensure a quality of life. A national studies suggest that technology may confer unique benefits in reducing difficulty with daily tasks and unmet need[42–46] and reduces the amount of personal care needed.[47] However, it remains unclear whether AT replaces or supplements personal care.

There is limited research on the substitutability of devices or technologi-cal assistance for personal care. Moreover, devices can substitute personal care (formal or informal care) in a number of ways. For example, a walking stick might alleviate the need for personal assistance and allowing an individual to move more independently. Adding grab-bars to the shower may enable older persons to shower independently and safely, making them more autonomous and reducing the need for human assistance. A descriptive research study in the United States,[37] found that simple assistive technology, such as canes, had the potential to substitute for informal care, while more complex devices, such as wheelchairs, appeared to supplement formal care. These relationships held after controlling for the underlying degree of disability severity.

In community-based long-term care studies there is a substitution be-tween technological and personal assistance.[26,37,48,49] Most research on long-term care in the community has evaluated the options in terms of formal and informal caregivers. The increased use of devices in home-based, long-term care arrangements has the potential to foster the independence and autonomy of older adults with ADL disability, alleviate the excess burden on caregiver, and reduce expenditure on long term care.

Older adults with ADL disabilities living at home or in the community compensate for their functional limitation through personal assistance and technological assistance. With the increasing demands for long-term care services for older adults, policymakers and researchers are interested in whether technological assistance can be a substitute for personal assistance or whether it is a complementary support. If this technological assistance is a substitute for personal assistance, such as when a cane is used rather than dependence on a human caregiver, then cost savings may be possible by promoting its use. Alternatively, if assistive technology is a complement, meaning both are used together, such as when a mechanical lifting device issued by a personal assistant for transferring a person to and from bed, then providing such technology will increase costs to the extent that it is provided, although it may reduce unmet need.[43] Limited research literature is available on this point, with some studies showing they are complementary,[50] some substitutes.[26,37,51] and some both[48,52] The differences across studies seem to depend on contextual factors, data, and analysis methodologies. Another important issue for understanding the relationship between technological assistance and personal assistance is controlling for differences in disability levels of recipients using these services. If the underlying need is relatively light, some assistive technologies may accommodate most or all of one's need for help. On the other hand, if underlying need is great, then substantial amounts of assistive technologies and human personal assistance may be needed. Controlling for differences in analyses will be important to accurately assess whether there is a substitution effect between technological assistance and personal assistance.[53]

References

1. Corbetta D, Imeri F, Gatti R. Rehabilitation that incorporates virtual reality is more effective than standard rehabilitation for improving walking speed, balance and mobility after stroke: a systematic review. *J Physiother.* 2015;61(3):117–124.
2. Vorrink SN, Kort HS, Troosters T, Lammers JWJ. A mobile phone app to stimulate daily physical activity in patients with chronic obstructive pulmonary disease: development, feasibility, and pilot studies. *JMIR Mhealth Uhealth.* 2016;4(1):e4741.
3. Hinman RS, Nelligan RK, Bennell KL, Delany C. "Sounds a bit crazy, but it was almost more personal:" a qualitative study of patient and clinician experiences of physical therapist–prescribed exercise for knee osteoarthritis via skype. *Arthritis Care Res (Hoboken).* 2017;69(12):1834–1844.
4. Ferorelli D, Nardelli L, Spagnolo L, et al. Medical Legal Aspects of Telemedicine in Italy: application Fields, Professional Liability and Focus on Care Services During the COVID-19 Health Emergency. *J Prim Care Community Health.* 2020;11.
5. Theodoros D, Russell T, Latifi R. Telerehabilitation: current perspectives. *Stud Health Technol Inform.* 2008;131:191–210.
6. Fairman A, Brickner A, Lieberman D, et al. Telerehabilitation. *Am J Occup Ther.* 2010;64(6):S92.
7. Torsney K. Advantages and disadvantages of telerehabilitation for persons with neurological disabilities. *NeuroRehabilitation.* 2003;18(2):183–185.
8. McLendon K. E-commerce and HIM: ready or not, here it comes. *J Am Health Inf Manag Assoc.* 2000;71(1):22–23.
9. Lee R, Conley D, Preikschat A. eHealth 2000: healthcare and the Internet in the new millennium. *Wit Capital.* 2000, January 31 Retrieved September 5, 2000.
10. Centers for Medicare and Medicaid Services Medicare Telemedicine Health Care Provider Fact Sheet. [(accessed on 15 September 2021)];2020. Available online: https://www.cms.gov/newsroom/fact-sheets/medicare-telemedicine-health-care-provider-fact-sheet.
11. Gil Membrado C, Barrios V, Cosín-Sales J, Gámez JM. Telemedicine, ethics, and law in times of COVID-19. A look towards the future. *Rev Clin Esp.* 2021;221:408–410. doi:10.1016/j.rce.2021.03.002.
12. Nittari G, Khuman R, Baldoni S, et al. Telemedicine Practice: review of the Current Ethical and Legal Challenges. *Telemed J Health.* 2020;26:1427–1437.
13. Curfman A, McSwain SD, Chuo J, et al. Pediatric telehealth in the COVID-19 pandemic era and beyond. *Pediatrics.* 2021;148. doi:10.1542/peds.2020-047795.
14. Kaplan B. Revisiting health information technology ethical, legal, and social issues and evaluation: telehealth/telemedicine and COVID-19. *Int J Med Inform.* 2020;143.
15. Battineni G, Nittari G, Sirignano A, Amenta F. Are telemedicine systems effective healthcare solutions during the COVID-19 pandemic? *J Taibah Univ Med Sci.* 2021;16:305–306.
16. Hsu C-H. Ubiquitous intelligence and computing: building smart environment in real and cyber space. *J Ambient Intell Humaniz Comput.* 2012;3(2):83–85.
17. Wichert R, Mand B, Eds. (2017). Ambient assisted living - 9. AAL-Kongress, Frankfurt/M, Germany, April 20–21, 2016 (1st ed.). Springer International Publishing.

18. Sugano S, Shirai Y. Robot design and environment design – Waseda robot-house project. Proceedings of 2006 SICE-ICASE international joint conference; 2006.
19. Zhu N, Diethe T, Camplani M, et al. Bridging e-health and the internet of things: the sphere project. *IEEE Intell Syst.* 2015;30(4):39–46.
20. Williams F, Moghaddam N, Ramsden S, De Boos D. Interventions for reducing levels of burden amongst informal carers of persons with dementia in the community. A systematic review and meta-analysis of randomised controlled trials. *Aging Ment Health.* 2019 Dec 2;23(12):1629–1642.
21. Sri V, Jenkinson C, Peters M. Carers using assistive technology in dementia care: an explanatory sequential mixed methods study. *BMC Geriatr.* 2022;16.
22. Niezgoda JA, Mendez-Eastman S. The effective management of pressure ulcers. *Adv Skin Wound Care.* 2006;19(1):3–15.
23. Rathore DK, Srivastava P, Pandey S, Jaiswal S. A novel multipurpose smart wheelchair. *IEEE Students' Conf Elect Electron Comput Sci.* Mar 2014:14.
24. Leishman F, Monfort V, Horn O, Bourhis G. Driving assistance by deictic control for a smart wheelchair: the assessment issue. *IEEE Trans Human-Mach Syst.* Feb 2014;44(1):66–77.
25. Edwards NI, Jones DA. Ownership and use of assistive devices amongst older people in the community. *Age Ageing.* 1998 Jul 1;27(4):463–468.
26. Hoenig H, Taylor Jr DH, Sloan FA. Does assistive technology substitute for personal assistance among the disabled elderly? *Am J Public Health.* Feb 2003;93(2):330–337.
27. Tinker A, Lansley P Introducing assistive technology into the existing homes of older people: feasibility, acceptability, costs and outcomes. J Telemed Telecare. Jul 2005;11(1_suppl):1-3.
28. Bryant BR, Seay PC. The technology-related assistance to individuals with disabilities act: Relevance to individuals with learning disabilities and their advocates. *J Learn Disabil.* Jan 1998;31(1):4–15.
29. Cornman JC, Freedman VA, Agree EM. Measurement of assistive device use: Implications for estimates of device use and disability in late life. *Gerontologist.* Jun 2005;45(3):347–358.
30. Kylberg M, Löfqvist C, Horstmann V, Iwarsson S. The use of assistive devices and change in use during the ageing process among very old Swedish people. *Disability Rehab: Assistive Technol.* 2013 Jan 1;8(1):58–66.
31. Verbrugge LM, Rennert C, Madans JH. The great efficacy of personal and equipment assistance in reducing disability. *Am J Public Health.* Mar 1997;87(3):384–392.
32. Lubinski R, Higginbotham D. *Communication technologies for the elderly.* 1st ed. San Diego: Singular Pub Group; 1997.
33. Scherer MJ. *Living in the state of stuck: How assistive technology impacts the lives of people with disabilities.* Brookline Books; 2005.
34. Gitlin, LN. Assistive technology in the home and community for older people: Psychological and social considerations, 2002.
35. Manton KG, Corder L, Stallard E. Changes in the use of personal assistance and special equipment from 1982 to 1989: results from the 1982 and 1989 NLTCS. *Gerontologist.* 1993;33(2):168–176.
36. Norburn J, Bernard S, Konrad T, Woomert A, Defriese G, Kalsbeek W, et al. Self-care and assistance from others in coping with functional status limitations among a national sample of older adults. *J Gerontol Series B: Psychol Sci Social Sci.* 1995;50B(2):S101–S109.
37. Agree E, Freedman V. Incorporating assistive devices into community-based long-term care. *J Aging Health.* 2000;12(3):426–450.
38. Gitlin LN, Levine RE. Prescribing adaptive devices to the elderly: principles for treatment in the home. *Int J Technol Aging.* 1992.
39. Kohn JG, Leblanc M, Mortola P. Measuring quality and performance of assistive technology: Results of a prospective monitoring program. *Assist Technol.* 1994;6(2):120–125.

40. Sanford J, Megrew M. An evaluation of grab bars to meet the needs of elderly people. *Assist Technol.* 1995;7(1):36–47.
41. Steinfeld E, Shea SM. *Enabling home environments: Strategies for aging in place.* Department of Planning, School of Architecture and Planning, State University of New York at Buffalo; 1995.
42. Agree E. The influence of personal care and assistive devices on the measurement of disability. *Soc Sci Med.* 1999;48(4):427–443.
43. Agree E, Freedman V. A comparison of assistive technology and personal care in alleviating disability and unmet need. *Gerontologist.* 2003;43(3):335–344.
44. Taylor D, Hoenig H. The effect of equipment usage and residual task difficulty on use of personal assistance, days in bed, and nursing home placement. *J Am Geriatr Soc.* 2004;52(1):72–79.
45. Verbrugge L, Rennert C, Madans J. The great efficacy of personal and equipment assistance in reducing disability. *Am J Public Health.* 1997;87(3):384–392.
46. Verbrugge LM, Sevak P. Use, type, and efficacy of assistance for disability. *J Gerontol Series B: Psychol Sci Social Sci.* 2002;57(6):S366–S379.
47. Mann WC, Ottenbacher KJ, Fraas L, Tomita M, Granger CV. Effectiveness of assistive technology and environmental interventions in maintaining independence and reducing home care costs for the frail elderly: A randomized controlled trial. *Arch Fam Med.* 1999;8(3):210.
48. Allen S, Foster A, Berg K. Receiving Help at Home: The interplay of human and technological assistance. *J Gerontol Series B: Psychol Sci Social Sci.* 2001;56(6):S374–S382.
49. de Klerk MM, Huijsman R. Effects of technical aids on the utilization of professional care. A study among single 75-year olds. *Tijdschriftvoorgerontologie en geriatrie.* 1996;27(3):105–114.
50. Agree E, Freedman V, Cornman J, Wolf D, Marcotte J. Reconsidering substitution in long-term care: when does assistive technology take the place of personal care? *J Gerontol Series B: Psychol Sci Social Sci.* 2005;60(5):S272–S280.
51. Mortenson W, Demers L, Fuhrer M, Jutai J, Lenker J, Deruyter F. How assistive technology use by individuals with disabilities impacts their caregivers. *Am J Phys Med Rehabil.* 2012;91(11):984–998.
52. Allen S, Resnik L, Roy J. Promoting independence for wheelchair users: the role of home accommodations. *Gerontologist.* 2006;46(1):115–123.
53. Kaye HS, Kang T, Laplante MP *Mobility device use in the United States.* Vol. 14. Washington, DC: National Institute on Disability and Rehabilitation Research, US Department of Education; 2000.

Index

Page numbers followed by "*f*" and "*t*" indicate, figures and tables respectively.

CPI Antony Rowe
Eastbourne, UK
September 01, 2023